Life

Also by John Brockman

As Author

By the Late John Brockman
37
Afterwords
The Third Culture: Beyond the Scientific Revolution
Digerati

As Editor

About Bateson
Speculations
Doing Science
Ways of Knowing
Creativity
The Greatest Innovations of the Past 2,000 Years
The Next Fifty Years
The New Humanists
Curious Minds
What We Believe but Cannot Prove
My Einstein
Intelligent Thought
What Is Your Dangerous Idea?
What Are You Optimistic About?
What Have You Changed Your Mind About?
This Will Change Everything
Is the Internet Changing the Way You Think?
Culture
The Mind
This Will Make You Smarter
This Explains Everything
Thinking
What Should We Be Worried About?
The Universe
This Idea Must Die
What to Think About Machines That Think

Life

The Leading Edge of Evolutionary Biology, Genetics, Anthropology, and Environmental Science

EDITED BY

John Brockman

HARPER PERENNIAL

NEW YORK • LONDON • TORONTO • SYDNEY • NEW DELHI • AUCKLAND

HARPER PERENNIAL

HarperCollins books may be purchased for educational, business, or sales promotional use. For information please e-mail the Special Markets Department at SPsales@harpercollins.com.

FIRST EDITION

Designed by

Library of Congress Cataloging-in-Publication Data has been applied for.

ISBN 978-0-06-229605-4 (pbk.)

16 17 18 19 20 RRD 10 9 8 7 6 5 4 3 2 1

dedication here

Contents

Introduction

Life is the fifth volume in The Best of *Edge* series, following *Mind*, *Culture*, *Thinking*, and *The Universe*. Here we present eighteen pieces from the pages of *Edge.org*, an online salon consisting of interviews, commissioned essays, and transcribed talks, many of them accompanied with streaming video and available, gratis, to the general public.

Edge, at its core, consists of the scientists, artists, philosophers, technologists, and entrepreneurs at the center of today's intellectual, scientific, and technological landscape. Through its lectures, master classes, and annual dinners in California, London, Paris, and New York, *Edge* assembles the thinkers who are exploring and rewriting our global culture.

Edge.org was launched in 1996 as the online version of the Reality Club, an informal gathering that met from 1981 to 1996 in Chinese restaurants, artists' lofts, the boardrooms of Rockefeller University and the New York Academy of Sciences, investment banking firms, ballrooms, museums, living rooms, and elsewhere. Though the venue is now in cyberspace, the spirit of the Reality Club lives on, in lively back-and-forth conversations on hot-button ideas. In the words of novelist Ian McEwan, a sometime contributor, *Edge.org* is "openminded, free ranging, intellectually playful . . . an unadorned pleasure in curiosity, a collective expression of wonder at the living and inanimate world . . . an ongoing and thrilling colloquium."

This is science set out in the largely informal style of a conversation among peers—nontechnical and colloquial, in the true spirit of the "third culture," which I have described as consisting of "those scientists and other thinkers in the empirical world who, through their work and expository writing, are taking the place of the traditional intellectual in rendering visible the deeper meanings of our lives. . . ."

Like the Modern Evolutionary Synthesis that revolutionized biology in the mid-20th century by marrying Mendelian genetics to Darwinian natural selection, the rise of biotechnology—epitomized by the Human Genome Project, which began in the 20th century's last decade and was completed in 2003—constitutes a turning point in our perception of who we are and where we're headed. The achievements of the past two decades have apparently made it possible for us—even incumbent upon us—not just to ameliorate the ills besetting our planet but to begin participating in our own evolution.

We have assembled contributions from some of *Edge*'s best minds, among them geneticists, theoretical biologists, theoretical physicists, and bioengineers. Pioneers in the current revolution, they take us over the work leading up to and in the wake of the Human Genome Project, illuminating in the process the conversations and controversies that animate modern biology.

We lead off with a 2015 talk by the evolutionary biologist **Richard Dawkins**, who defends his "selfish gene" view of Darwinian natural selection and speculates that life elsewhere in the universe will exhibit it. Evolutionary geneticist and theorist **David Haig** follows, with a talk on conflicts and conflict resolution within the genome, stemming from maternal and paternal imprinting. **Robert Trivers**, canny lone wolf of evolutionary biology, discusses self-deception and the "biased information flow between the conscious and the unconscious." The late **Ernst Mayr**, an architect of the 20th century's modern synthesis, talks to *Edge* about the course of evolutionary biology since then and his agreements and disagreements with it. The highly regarded geneticist and snail biologist Steve Jones comments on the robustness of Darwin's 150-year old worldview.

E. O. Wilson recalls the the divide (lately bridged) at Harvard between the morphologist/naturalists and the newly arrived molecular biologists—particularly personified by the rambunctious Jim Watson. Theoretical physicist **Freeman Dyson** speculates about the future of life—will it be analog or digital? Then, at an *Edge* gathering at Eastover Farm in Connecticut a few years later, he joins biotechnologist and entrepreneur **J. Craig Venter**, geneticist **George Church**, astrophysicist **Dimitar Sasselov**, and quantum engineer **Seth Lloyd** in a free-ranging discussion of the origins of life and its prospects, here on Earth and elsewhere. This is the book's longest section and its centerpiece; in it, *Life's* major themes and arguments play out, sometimes brilliantly, sometimes with a lot of tongue in cheek. A year later, Dawkins and Venter renew the discussion in a dialogue on their pet theories, enlivened by the issues that divide them.

Armand Marie Leroi, a developmental biologist at Imperial College London, ponders the wide range of genetic variation in the human species, along with the reluctance of some scientists to study such matters as skin color "because of the long and sorry history of genetics and racial differences." The physiological legacy of our hominid ancestors is discussed by paleoanthropologist **Daniel Lieberman**, and our possible Neanderthal heritage by **Svante Pääbo**, mapper of the Neanderthal genome. Craig Venter is joined by inventor and futurist **Ray Kurzweil** and roboticist **Rodney Brooks** in a session on recent advances in genomics and biotechnology; bioengineer **Drew Endy** makes a case for focusing on the engineering aspects of synthetic biology.

Kary Mullis, inventor of the polymerase chain reaction

(PCR) in the 1980s, which enabled the sequencing and cloning of DNA, talks about his current attempts to improve the human immune system. Yale evolutionary ornithologist **Richard Prum** talks about the importance of aesthetics in natural selection, reviving the old argument between Darwin and Alfred Russel Wallace, the co-discoverer of evolution by natural selection. Along the way, he tells you more than you might want to know about the love life of ducks. Neuroendocrinologist **Robert Sapolsky** warns us about the ubiquity and ingenuity of the parasite protozoan *Toxoplasma*; and the theoretical biologist **Stuart Kauffman**, an expert on complex systems, concludes the collection with an essay on how the universe got complex and on whether or not there are "laws that govern biospheres across the universe."

Life—and particularly intelligent life—has been called an "emergent" phenomenon. But has it fully emerged? Now that we are a party to its "emergence," what are the opportunities before us, and what are our responsibilities with regard to a continuing and partly man-made evolution—on (and perhaps someday off) this planet? Ought we to "play God," as some naysayers have vigorously accused the 21st century's geneticists and biotechnologists of doing—or are we simply meeting our obligations and living up to our natural potential as human beings?

John Brockman
Editor and Publisher
Edge.org

1
Evolvability

Richard Dawkins

[April 30, 2015]

Richard Dawkins **is an evolutionary biologist and Charles Simonyi Professor of the Public Understanding of Science, Emeritus, at Oxford.**

Natural selection is about the differential survival of coded information which has the power to influence its probability of being replicated, which pretty much means genes. Whenever coded information which has the power to make copies of itself—a "replicator"—comes into existence in the universe, it potentially could be the basis for some kind of Darwinian selection. And when that happens, you then have the opportunity for this extraordinary phenomenon we call "life".

My conjecture is that if there is life elsewhere in the universe, it will be Darwinian life. I think there's only one way for this hypercomplex phenomenon we call life to arise from the laws of physics. The laws of physics—if you throw a stone up in the air, it describes a parabola, and that's it. But biology, without ever violating the laws of physics, does the most extraordinary things; it produces machines which can run, and walk, and fly, and dig, and swing through the trees, and think, and produce the whole of human technology, human art, human music. This all comes about because at some point in history, about 4 billion years ago, a replicating entity arose—not a gene as we would now see it, but something functionally equivalent to a gene— which, because it had the power to replicate and the power to influence its own probability of replicating, and replicated with slight errors, gave rise to the whole of life.

If you ask me what my ambition would be, it would be that everybody would understand what an extraordinary, remarkable thing it is that they exist, in a world which would otherwise just be plain physics. The key to the process is self-replication. The key to the process is that—let's call them "genes," because nowadays they pretty much all are genes—genes have different probabilities of surviving. The ones that survive, because they have such high-fidelity replication, are the ones we see in the world, the ones which dominate gene pools in the world. For me, the replicator—the gene, DNA—is absolutely key to the whole process of Darwinian natural selection. So when you ask the question "What about group selection, what about higher levels of selection, what about different levels of selection?", everything comes down to gene selection. Gene selection is fundamentally what's really going on.

Originally, these replicating entities would have been floating free and just replicating in the primeval soup, whatever that was. But they "discovered" a technique of ganging together into huge robot vehicles we call individual organisms. An individual organism is a unit of selection in a different sense from the replicator being a unit of selection. The replicator is the unit of selection which strictly is the thing that becomes either more numerous or less numerous in the world. Nowadays we say more numerous or less numerous in the gene pool, and that's modern post-Darwin language.

But because the individual organism is such a salient unit in which these replicators, these genes, have ganged up together, we as biologists tend to see the individual organism as the unit of action. The individual organism is the thing that has legs or wings, it has eyes, it has teeth, it has instincts. It's the thing that actually does something. And so it's natural for biologists to phrase their questions of purpose, of pseudo-purpose, at the level of the organism. They see the organism as striving for something, working for something, struggling to achieve something.

What's it struggling to achieve? Well, for Darwin it was struggling to achieve survival and reproduction. Nowadays we would say it's struggling to achieve replication of the genes inside it. And this all comes about because, well, one way of putting it, and I've often put it like this, is to say, "Look backward at the ancestors of all modern animals," any animals, anytime, and you can see that the individual is descended from an unbroken line of successful ancestors, an unbroken line of individuals who succeeded in surviving and reproducing. What that really means is that they succeeded in passing on the genes that built them. So we are conduits for the genes that pass through us. We are temporary survival machines.

Everything about biology can be understood in this way. Everything about biology can be understood if you say that what's really going on is differential replicator survival—gene survival in gene pools—and the way in which they do it is by controlling phenotypes. And those phenotypes in practice are nearly all bundled up into these discrete bodies, individual organisms.

If ever there is a bundle of replicators, a bundle of genes, which passes on its genes to the next generation in a single propagule (we do that: We pass on our genes in sperms or eggs), that means that all the genes in a body—in a mammal body, in a vertebrate body, in an animal body, a normal animal with sexual reproduction—have the identical expectation of getting into future generations: namely, leaving the present body in a sperm or an egg. That means that all the genes in a body are pulling for the same end. They all have the same goal.

If they didn't (and some of them might not: Viruses, for example, have a different goal, of being sneezed out or spat out or whatever it might be), they of course are quite different, and they do not cooperate with the rest of the genes in the body. But all the genes that have the same expectation of the future, the same expectation of leaving the present body and getting into the next body, cooperate. They work together. That's why bodies are such coherent wholes. That's why all the limbs and all the sense organs work together. It's simply because all the genes that built them have the same exit route to the next generation. The minority that don't—things like viruses—have a different exit route, and they don't cooperate, and they may kill you.

Although it's true that the great majority of survival machines are discrete organisms, that doesn't necessarily have to be the case, and if genes can influence phenotypes outside the body, then they will do so. This is the extended phenotype. The simplest sort of extended phenotype would be an artifact, like a bird's nest. So a bird's nest is an organ. It's an organ in the same sense as a heart or a kidney is an organ, but it just happens to be outside the body and it happens to be made of grass and sticks rather than being made of the cells that contain the genes. Nevertheless, it's a phenotype, which is produced by the animal's nervous system working through nest-building behavior. And it does exactly the same kind of thing—namely, preserve the genes in the form of eggs and chicks, as organs of the body, like kidneys and livers and muscles.

The next kind of extended phenotype I talk about is hosts of parasites, because there are spectacular examples. For example, parasites which influence their hosts in order to get into the next host. A host body to a parasite gene is like a bird's nest: It's influenced by the genes. We don't normally put it that way; we normally say that the parasite, the fluke, or whatever it is, the whole fluke influences the whole snail, to get itself passed on. But if you think at the genetic level, the genes are influencing the fluke's phenotype, which, in turn, influences the snail's phenotype to enhance the propagation of the fluke's genes into the next generation. So there's no reason to draw a line around the fluke's body and say, "Well, outside that is no longer proper phenotype." It *is* proper phenotype, it's just that you have to think outside the box—in this case, outside the fluke—in order to get the true relationship between genes and phenotypes.

And then, generalizing further: A cuckoo in a nest influences the behavior of its host by various stimuli—by having a bright red beak and squawking in the right way and so on. And once again, just as the fluke influences the snail to get itself passed on to the next generation, the cuckoo influences the reed warbler to get itself, to get its genes, passed on to the next generation. And the change in reed-warbler behavior can properly be regarded as a phenotypic expression of cuckoo genes.

My vision of life on this planet is that everything extends from replicators, which are in practice DNA molecules. The replicators reach out into the world to influence their own probability of being passed on. Mostly they don't reach farther than the individual body in which they sit, but that's a matter of practice, not a matter of principle. The individual organism can be defined as that set of phenotypic products which have a single route of exit of the genes into the future. That's not true of the cuckoo / reed warbler case, but it is true of ordinary animal bodies. So the organism, the individual organism, is a deeply salient unit. It's a unit of selection in the sense that I call a "vehicle".

There are two kinds of unit of selection. The difference is a semantic one. They're both units of selection, but one is the replicator, and what it does is get itself copied, so more and more copies of itself go into the world. The other kind of unit is the vehicle. It doesn't get itself copied. What it does is work to copy the replicators which have come down to it through the generations, and which it's going to pass on to future generations. So we have this individual / replicator dichotomy. They're both units of selection, but in different senses. It's important to understand that they are different senses.

Now, because the individual organism is such a salient unit, biologists after Darwin got into the habit of seeing the organism as the unit of action, and therefore they asked the question, "What is the organism maximizing?" What mathematical function is the organism maximizing? "Fitness" is the answer. So fitness was coined as a mathematical expression of that which the organism is maximizing. Of course, what fitness really is, or what it ought to be if we understand it properly, is gene survival. For a long time, fitness was equated in people's minds with reproduction, with having a large number of children, grandchildren, great-grandchildren. Bill Hamilton and others, but mostly Bill Hamilton, realized that you had to generalize that, because if what's really going on is working to pass on genes, then offspring—grandchildren, et cetera—are not the only ways of passing on genes. An organism can work to enhance the survival and reproduction of its siblings, its nephews, its nieces, its cousins, and so on. Hamilton worked out the mathematics of that.

It was unfortunate that Hamilton, having realized this important insight, chose to stick with the individual organism as the entity of action. He therefore coined the phrase "inclusive fitness" as the mathematical function an individual organism will maximize if what it's really doing is maximizing its gene survival. It's a rather complicated thing to calculate. It's difficult to calculate in practice, and this has led to a certain amount of—not hostility, but a certain amount of skepticism about inclusive fitness as a measure, skepticism I share. But for me, the remedy for that skepticism is to say, "Well, forget about the organism and concentrate on the gene itself." Ask yourself, as Hamilton also did, "If I were a gene, what would I do to maximize my propagation into the future?" Hamilton did that, but he also later took a sort of false trail (it's strictly correct but not helpful) by asking, "If I'm an individual, what would I do to maximize my gene survival?" Both ways of phrasing it are correct—they're both correct if you can get the calculation right—but one of them is rather harder to do. If you're trying to do intuitive Darwinism, if you're trying to work out what you would expect to happen in the world, I think it's better to ask the question "What would I do if I were a gene?" rather than "What would I do if I were an elephant?"

In both cases this is a personification. Nobody really thinks that either genes or elephants scratch their heads and think, "What would I do?" But it's a useful trick, a useful dodge when you're trying to get the right answer as a field biologist in the Serengeti. It's a useful trick to say what would I do if I was a . . . and you could fill in the end of that sentence by saying either "if I was a gene" or "if I was an elephant." And you'll get the right answer if in the gene case you concentrate on self-replication and if in the elephant case you concentrate on passing on genes. So we have these two logically equivalent ways of expressing what's going on in Darwinism. Both of them Hamilton used. I think some of the opposition to Hamilton, which has recently surfaced, is because people have realized that inclusive fitness is not a practical way of doing things. It's a difficult thing to calculate. And my suggestion would be—and I said this to Hamilton—to abandon inclusive fitness and concentrate instead on personification of the gene, and then you'll get the right answer.

George C. Williams in 1966 wrote a brilliant book, *Adaptation and Natural Selection*, roughly at the same time as Hamilton was working, and they both tumbled to the same truth, which is that what's really going on in natural selection is survival of genes. Williams was eloquent on this. Williams said things like, Socrates may have had any number of children, we don't know that, but what Socrates really passed on, if he passed on anything, was genes. It's genes that pass through the generations. And so whenever you're talking about teleonomy, whenever you're talking about pseudo-purpose, which is what we see in life—what's it for, what's the adaptation for, who benefits, *cui bono*—whenever you ask that question, you should be looking at the level of the gene. Williams realized that; Hamilton realized that.

In *The Blind Watchmaker,* I wanted to get across the idea that cumulative selection can give rise to immense complexity and dramatic changes. So I wrote a computer program for the Macintosh, which presented on the screen a range of phenotypes built by an algorithm I called its embryology, which was actually a tree-growing algorithm. And the shape of the tree was governed by genes. There were 9 genes, I think, in the first version, and so what the user saw on the screen was a "parent", as I called them, in the middle, and 14 other biomorphs around it were the offspring. They were built by genes, which were 9 numbers. The genes could mutate by either having a small amount added to their value or a small amount subtracted from their value. So all the 9 biomorphs looked a bit different—obviously descended from the same parent but they were a little bit different. And you could choose with a mouse which one to breed from; it glided to the center of the screen, produced 14 offspring and so on. It went on and on through generation after generation. You could breed anything you liked. It was a most extraordinary experience, to breed massively different shapes from the original by gradual degrees, and they came out looking like insects, and flowers, and all sorts of things.

I'm pleased to note that although I'd thought I'd lost these biomorphs—because modern Macs don't run the software that old Macs do—a wonderful man called Alan Canon in Kentucky wrote to me and said he wanted to revive them. So I sent him all my old Pascal code, which would no longer run, and he's now hard at work producing phoenix from the ashes—my old programs—and I'm simply delighted by this.

I then went to the Artificial Life Conference organized by Chris Langton, and I gave a talk called "The Evolution of Evolvability," which I think was the first time the phrase had ever been used, and it's being used quite a lot.

The original biomorph program had 9 genes. I later enlarged it to 16 genes. I added genes that did things like segmentation, that had biomorphs arranged serially along the body, like a centipede, which has lots of different segments, or a lobster, which has lots of segments but each segment can be a little bit different. I had genes that had symmetries of various kinds. So the repertoire of biomorphs that became possible to breed then dramatically increased. It was still limited, but nevertheless it increased. And it occurred to me that this was a good metaphor for radical changes in embryology that happened at certain important times in evolution. For example, I just mentioned segmentation. The very first segmented animal had some kind of major mutation which gave it two segments instead of one, I'm guessing. It may have been three. It can't have had just one-and-a-half segments; there must have been at least two. It duplicated everything about the body. If you look at the body of an earthworm or a centipede, it's like a train, like a truck. Each truck is similar to the neighboring trucks and may be identical.

Before the origin of segmentation in the ancestors of earthworms or centipedes, the ancestors of vertebrates, animals must have evolved as one single segment, and they would have evolved in the same sort of way as my biomorphs did when they had only 9 genes. Then the first segmented animal was born. It must have been radically different from its parents. This must have been a major mutation. And as soon as the first segmented animal was born with two segments—the same as each other, probably—it wasn't a difficult thing to do in one sense, because all the embryological machinery to make one segment was already there. To double it would have been a major step; nevertheless all the machinery is there. It's not like inventing a whole new organ, like an eye. That cannot happen. It's got to happen by gradual cumulative selection, which is the main message of *The Blind Watchmaker*. But once you've got the machinery to make an eye, or to make a vertebra, or to make a heart or anything like that, you could make two, because the machinery is already there. That's what segmentation is.

And so when segmentation was invented by some kind of macro mutation, a whole new flowering of evolution became possible, and vertebrates, arthropods, annelids, all exploit this new embryological trick of segmentation. I illustrated this with my biomorphs, because when I added the segmentation gene for the macro mutation, which I had to program in, a whole new flowering of morphology could appear on the screen. You could evolve much more exciting animals because segmentation was there. Similarly with the genes for symmetry. I had genes doing kind of mirror-image morphology in two different planes. And immediately I started being able to breed things like flowers, butterflies, beautiful creatures.

The evolution of evolvability, then, is an evolutionary change which makes a radical alteration in embryology and opens up floodgates of further evolution not possible before. Segmentation is one example, sex may be another one. Torsion in mollusks may be another one. These are major changes, which I think are rare. They may happen once every 100 million years. There's normal evolution, which goes on by the normal cumulative, gradual process we mostly teach about. But every now and again, I suspect there's a major jump, a macro mutation which opens up new floodgates, and segmentation would be the best example. I was led to think about this by the addition of 7 more genes to my original 9-gene biomorph, and that's what I talked about at Chris Langton's Artificial Life Conference, and I called it "The Evolution of Evolvability."

I incorporated these ideas of evolution of evolvability in *Climbing Mount Improbable*, which is a bit similar to *The Blind Watchmaker* but has a lot more in it. And by then I'd added a whole lot more genes, in this case introducing colors, and we now have color biomorphs. And perhaps rather more interesting, I teamed up with Ted Kaehler, one of Apple's star programmers. I met him at the Artificial Life Conference. And after that we collaborated on a new project I called Arthromorphs, which was somewhat similar to biomorphs but with a totally different kind of embryology and segmentation and much more based upon especially arthropod segmentation. The arthromorph program didn't require the programmer—namely, me—to introduce the new watershed changes, the new macro mutations which led to new flowerings of evolution. It happened internally; it happened in the computer. They really were macro mutations. That was a big step in my use of computers in both understanding and teaching about evolution.

One of the things I've always done is not make a clear separation between books aimed at popularizing, books aimed at explaining things to people, and books that explain things to myself or my scientific colleagues. I think the separation between *doing* science and *popularizing* science has been overdone. I have found that the exercise of explaining to other people, which I suppose I've been fairly successful at, is greatly helped by the fact that I first have to explain it to myself. And explaining it to myself—the biomorph program, which I originally wrote to explain to students, and I used them in student practicals—led me to think anew for myself, stimulated me to understand much better about evolution, stimulated me to understand about the evolution of evolvability in a way I haven't before.

Nobody knows whether there's life elsewhere in the universe. I think there probably is. The number of stars in the universe is something like 1022, and most of them have probably got planets. It would be pretty astonishing if we were unique. It would go against the lessons of history: You know, we're not the center of the universe, et cetera. Science fiction writers try to speculate about what life elsewhere might be like. I have one contribution to make, which is that I think however weird and alien and strange and different life elsewhere might be, we can say one thing about it, which is that, if we discover it, it will turn out to be Darwinian life.

I think there's only one way for the lead of pure physics to be transmuted into the gold of complex life, and that is differential replicator survival, which is Darwinism in its most general sense. So I would stick my neck out and say that when and if we ever discover life elsewhere in the universe, it will be Darwinian, it will be based upon something like DNA—probably not DNA, but something like DNA in the sense of an ultra-high-fidelity, self-replicating coding system able to produce great variety, which is what DNA does. So what I call universal Darwinism is a doctrine, almost: The one thing we know about life everywhere is that it's Darwinian life.

I gave a talk called "Universal Darwinism" at one of the Darwin Centenary Conferences, the one in Cambridge, and I based it on looking at all the alternatives that someone might have suggested, like Lamarckism—the inheritance of acquired characteristics, the principle of use and disuse. The point I tried to make is that contrary to what most biologists have said, the thing that's wrong with Lamarckism is not just that it doesn't work in practice, that acquired characteristics are not, as a matter of fact, inherited. There are biologists, including Ernst Mayr, who have said Lamarck's theory is a fine theory but, unfortunately, acquired characteristics are not inherited. The point I made was that even if they *were* inherited, the Lamarckian theory is nothing like a big enough theory to do the job of producing complex adaptations. Lamarckian theory depends on use and disuse: The more we use our muscles, the bigger they get. That's fine, that happens, and then—inheritance of acquired characteristics—you pass on your bigger muscles to your children. Ernst Mayr said that's a perfectly good theory and the only trouble is it doesn't work, because acquired characteristics are not inherited, which of course is true. But the point I was making was that even if it were true, the principle wouldn't work to produce real, interesting biological evolution.

Muscles are fine; that's one thing that does grow bigger when you use them. But something like an eye, the delicate focusing mechanism of the eye, the transparency of the eye, the huge number of light-sensitive cells, 3 different color codings and so on—that doesn't come about by use and disuse. The more you use your eyes, they don't become more— The lens doesn't become more transparent as photons wash through it. The eyes become better because of every single tiny mutation that improves the eye. As Darwin said, Nature is daily and hourly scrutinizing. So every little tiny change, no matter how deeply buried in internal cellular biochemistry it is, if it has any effect whatever on survival and reproduction, natural selection will pick it up. The Lamarckian principle will work only for very, very crude growth—things like muscles getting bigger when you use them.

As we look around the world in which we live, what we see is stupefyingly complicated man-made machines like this camera you're filming with, this recording machine, this computer, cars, ships, planes. These are not produced directly by natural selection, these are produced by human ingenuity, by human brains working together. No one human can make a Boeing 747. I mean, this is a cooperative enterprise involving lots of humans, involving lots of computers. It's a fantastic extension of the Darwinian substrate. So the principles that give rise to the strong design of a plane, or a car, or a computer—these all come from human brains. But that's not the ultimate explanation. The human brains themselves have to come from Darwinian natural selection. So if we go to other planets and discover extremely complicated technology, that technology itself will be the direct product of Darwinian selection, but it will be the product, ultimately, of Darwinian selection of the brains—or whatever they call them on that planet.

It's arguable that something—this is a different kind of argument now—it's arguable that something like Darwinism does go on in human technology: that when a human designer is designing on the drawing board, he designs something, doesn't like it, tosses it in the bin, gets a fresh bit of paper, designs a slight variation of it, and so on. There might be a Darwinian element to that. That's not what I'm saying. I'm saying that a wholly new, at least partly new, kind of design came into the world when human brains started to exercise ingenuity, especially social ingenuity, cultural ingenuity. But the ultimate source of that is evolved brains, and the evolved brains have to come about by some version of Darwinian selection, which on other planets might be very different, but it will still be—I conjecture, I bet my shirt on it—Darwinian.

2
Genomic Imprinting

David Haig

[October 22, 2002]

David Haig is a professor in Harvard's Department of Organismic and Evolutionary Biology.

My work over the last decade or so has been principally concerned with conflicts within the individual organism. In a lot of evolutionary biology, the implicit metaphor is that the organism is a machine or, more specifically, a fitness-maximizing computer trying to solve some problem. I'm interested in situations where there are conflicts within the individual, in which different agents within the self have different fitness functions—as well as the internal politics resulting from those conflicts of interest.

The area to which I've given the greatest attention is a new phenomenon in molecular biology called genomic imprinting, which is a situation in which a DNA sequence can have conditional behavior depending on whether it's maternally inherited from an egg or paternally inherited through a sperm. The phenomenon is called imprinting because the basic idea is that some imprint is put on the DNA in the mother's ovary or in the father's testes which marks that DNA as maternal or paternal and influences its pattern of expression—what the gene does in the next generation, in both male and female offspring.

This is a complicated process, because the imprint can be erased and reset. For example, the maternal genes in my body, when I pass them on to my children, will be paternal genes, having paternal behavior. If my daughter passes on paternal genes to her children, even though she got the gene as a paternal gene from me, it will be a maternal gene to her offspring. Molecular biologists are particularly interested in understanding the nature of these imprints, and how it is possible to modify DNA in some way that's heritable but can then be reset. My own interest has been in understanding why such odd behavior should evolve. I've been trying to find situations in which what is best for genes of maternal origin is different from what maximizes the fitness of genes of paternal origin.

The best way to understand the underlying theory is with a famous anecdote credited to J. B. S. Haldane, the great British geneticist, who is said to have claimed that he would give his life to save more than two drowning brothers or more than eight drowning cousins. The logic is that if Haldane is concerned only with transmitting his genes to future generations, this is the right thing to do. On average, a gene in his body has one chance in two of being present in a brother. If he sacrificed the copy of a gene in his body to rescue three brothers, on average he'd be rescuing one-and-a-half copies of the gene in his three brothers, placing him ahead in the genetic accounting. But when it comes to cousins, each only has one chance in eight of carrying a random gene in Haldane's body. To benefit from the sacrifice of one copy of a gene in himself, he needs to rescue nine or more cousins. This was formalized by Bill Hamilton in his theory of inclusive fitness.

My theory can be illustrated by rephrasing Haldane's question and asking: "Would Haldane sacrifice his life for three half-brothers?" For the sake of the story, let's say these are his maternal half-brothers—offspring of his mother but with different fathers. The traditional answer to that question is no, because if you pick a random gene in Haldane, it's got one chance in four of being present in a half-brother. Thus, a random gene would have an expectation of rescuing three-quarters of a copy—3 x 1/4—for the loss of one copy in Haldane. However, if imprinting is possible, genes may have information about their parental origin, and this can change the accounting.

From the point of view of a maternally derived gene in Haldane, the three half-brothers are all offspring of his mother, so his maternally derived genes have a probability of one-half being present in each half-brother. For the sacrifice of one copy of the gene in himself, Haldane would be rescuing one-and-a-half copies, on average, of his maternally derived genes. Natural selection acting in that situation on genes of maternal origin would favor the sacrificial behavior.

However, things look very different from the point of view of Haldane's paternal genes. Those three half-brothers are the offspring of different fathers, making them complete non-relatives. If genetic accounting were all that was important, no sacrifice, no matter how small, would justify any benefit, no matter how great, to his paternal half-sibs. Therefore, in this case, selection on paternally derived genes would prevent Haldane performing this sacrificial action.

This illustrates that different selective forces can act on different genes within an individual, pulling him in different directions, resulting in internal genetic conflicts. I suspect that how these conflicts are resolved is a matter of history, genetic politics, and knowing the details of the system. A lot of insight will come from the social sciences: Political science in particular is all about dealing with conflicts of interest within society with the formations of parties and factions, and I believe that if there are conflicts within the individual, you'll have a similar sort of internal politics.

I'm particularly interested in looking at situations in the real world where the Haldane story I just gave would apply—where there are potential conflicting selective forces acting within the individual. So far I've talked about conflicts between genes of maternal and paternal origin, but there are also possible conflicts between genes sitting on the sex chromosomes and genes sitting on the other chromosomes, and between genes sitting in the nucleus and genes sitting in mitochondria, and between our genetic inheritance and cultural transmission. I'm trying to develop a set of theories and tools for dealing with such situations.

Genomic imprinting is a fascinating phenomenon, and raises an interesting question: If information about the sex of the parent in the previous generation can be transmitted by such mechanisms, is there other historical information input from the environment that can be transmitted to the current generation and influence genetic expression? Would it be possible that if my great-grandmother experienced a famine or lived in a time of war, that this has put an imprint on the genome which is influencing gene expression in my own body?

My interest in genetic imprinting began while I was completing my doctorate at Macquarie University in Sydney. I began studying plant ecology and, in particular, how regeneration after fire takes place. I wandered around the bush looking at plants, but my heart really wasn't in that. Through good fortune, I got an opportunity to do a theoretical study on the evolution of the life cycles of plants, applying kin-selection theory—the theory of parent-offspring conflict developed by Robert Trivers—to plants. By thinking about what's happening within seeds, I essentially had a theory of genomic imprinting ready to go the moment I heard of the phenomenon.

In a 1974 paper on parent-offspring conflict [*American Zoologist*, 14:1, 249-64 (1974)], Trivers pointed out that there was often an implicit assumption that what was good for a parent was also good for the offspring. In terms of genetic transmission, it would seem that offspring are parents' stake in the future, so parents should be doing their best for them. What Trivers argued, however, was that parents would be selected to maximize their total number of surviving offspring—which may be quite different from maximizing the survival of an individual offspring. He suggested that there was a tradeoff between producing lots of offspring and investing relatively little in them versus producing a small number of offspring and investing a lot in each. He thought that over evolutionary time offspring would begin to compete with their siblings for available resources. And in turn, sibling rivalry would result in conflict between offspring and parents, since over time offspring would be selected to try to get more than their fair share of resources from their parents—more than the parents were selected to supply—whereas parents would be selected to spread their resources more evenly over a larger number of offspring. Trivers's theory was that this could lead to evolutionary conflicts.

I was asked to talk at the National Institutes of Health in a workshop on imprinting and human disease. My goal was to suggest how evolutionary theory would provide new insights into human disease. An obvious case was in human pregnancy, where Trivers's theory of parent-offspring conflict could help explain why pregnancy is so often associated with medical complications. Since then, looking at maternal fetal interactions has been another area in my research.

Trivers's theory has a lot to say about why pregnancy doesn't work particularly well. If we look at most of the products of natural selection—like the hand, the liver, the heart, or the kidney—these are wonderful bits of engineering that function well for 60 or 70 years. But why are there so many problems in pregnancy? Pregnancy is essential to reproduction, so you might expect that this would be one part of our human physiology that had been perfected by natural selection. But there's an important evolutionary difference between the function of the heart and what's going on in pregnancy. When we look at the selective forces acting on the function of the heart, there's no evolutionary conflict. All of the genes involved in the development and function of the heart belong to the same genetic individual and, in a sense, have the same genetic interest: the maximization of the number of offspring of that individual. In the absence of conflict, we've got a simple optimization problem, and you get an optimal solution.

But in the relationship between mother and fetus, because of the parent-offspring conflict that Trivers pointed out, we've now got conflicting forces. The offspring is being selected to take a little extra from the mother, and the mother is selected to resist some of the offspring's demands. Those selective forces tend to act at cross purposes and cancel each other out.

One important problem during pregnancy is the communication of information between mother and offspring. In communication within the body, there's no conflict, since selection causes cells to send messages as cheaply and efficiently as possible. But when you're looking at the exchange of messages between mother and fetus, there's a problem of credibility, since their interests are not identical. In some situations, there's an evolutionary incentive to send misleading messages, and corresponding selection for receivers to distrust messages being received.

One thing that's happening during pregnancy is that there's a lack of the usual feedback controls, checks and balances. I read grant applications for scientists proposing to study maternal-fetal relations, and they tend to portray it in rosy terms, as an almost loving exchange of messages between mother and fetus. But in pregnancy an embryo implants itself in the abdominal cavity or in the fallopian tube—in a completely inappropriate position in the body—and develops autonomously in the absence of any appropriate maternal messages. I believe there's very little communication going on between the mother and the fetus during pregnancy. Rather, you're looking at various fetal attempts to manipulate maternal physiology and metabolism for fetal benefits.

During pregnancy, the mother's hormonal communication systems are coming under joint control of both the mother and the fetus. The fetus secretes a number of hormones into the mother's body to achieve various effects, particularly increasing the nutrient levels of the maternal blood. In the early stages of human pregnancy, the embryo embeds itself in the uterine wall and taps into the maternal blood system, releasing hormones into maternal blood that can influence the mother's physiology, blood-sugar levels, and blood pressure. The higher the levels of sugar and fats in maternal blood, the more nutrients the fetus can obtain. Typically, hormones are molecules produced in tiny amounts that have big effects, at least when communication occurs within a single body and there's no conflict between sender and receiver. However, in pregnancy, one individual (the fetus) signals to another (the mother) and there is potential for conflict. Natural selection favors increased production of the hormones by offspring to get a bigger effect, while at the same time it favors maternal receiving systems that become more and more resistant to manipulation. There is thus potential for an evolutionary escalation that sometimes results in placental hormones being produced in massive amounts. It's estimated that about a gram a day of human placental lactogen is secreted into the maternal blood stream, and yet it has relatively minor effects.

I think this observation—that placental hormones tend to be produced in very large amounts—is the best evidence for the existence of maternal-fetal conflict. The fetus secretes these hormones into the mother's body in an attempt to persuade the mother to do something she might not necessarily want to do. Think of placental hormones as the equivalent of junk mail. These messages are trying to persuade you to do something. They're relatively cheap to produce, so they're distributed in vast quantities but have relatively minor effects. They must work sometimes, but it's very different from the sort of intimate whisper you might get between two individuals who have common interests.

The most successful application of my ideas on imprinting has been to the study of growth during pregnancy, and the prediction that paternally derived genes are selected to produce larger placentas that extract more resources from mothers. But the basic idea of the theory applies to any interactions among relatives that are what I call asymmetric kin—that is, relatives on the maternal side of the family but not on the paternal side, or vice versa. I suspect that genomic imprinting is going to be relevant to understanding the evolution of social interactions. There's also evidence now that imprinting is implicated in some forms of autism. There are a number of imprinted genes that are known to be imprinted in the brain, and I'm interested in exploring those ideas.

The most exciting empirical work that's been done to test my ideas came out of Shirley Tilghman's lab before she became president of Princeton. Hers was one of the first labs to describe an imprinted gene. Paul Vrana, a postdoc of Tilghman's, looked at crosses between two species of mice, one of which had a high rate of partner change—multiple fathers within a litter—whereas the other was a so-called monogamous mouse, where a single father fathered all the offspring in a litter and the female had about an 80-percent chance of staying with the father to produce the next litter. The researcher predicted that the conflict between maternal and paternal genomes would be more intense in the mouse with multiple paternity than in the monogamous mouse, and in fact when you cross them you get a dramatic difference in birth weight.

If the father came from the species with multiple paternity, there had been intense selection on paternal genomes to extract more resources from mothers. This paternal genome would be matched against a maternal genome that had not been strongly selected to resist paternal demands. In this direction of the cross, offspring were larger than normal, whereas in the reciprocal cross, where the paternal genome came from the monogamous species and the maternal genome from the polyandrous species, offspring were smaller than normal. Paul Vrana was able to show that this difference was largely due to imprinted genes in these two species. This suggests that divergence of imprinted genes may contribute to the speciation process, and in particular that changes in social systems and mating systems can cause changes in the expression of imprint. These can then contribute to reproductive isolation between sister species.

The second bit of work is being done in, of all places, a liver oncology lab at Duke University Medical Center which is studying genomic imprinting. Out of curiosity, Randy Jirtle and Keith Killian looked at marsupials and then at the platypus—an egg-laying mammal—to see where imprinting arose. They found that imprinting was absent in the platypus, at least for the genes they looked at, but was present in marsupials. Thus, imprinting appears to have arisen more or less coincident with the origin of live birth, before the common ancestor of marsupials and placental mammals. There are some exciting areas of research of that kind.

There are other recent intriguing observations out there that beg for a theoretical explanation. There's evidence in the mouse, for example, that the paternal genome particularly favors development of the hypothalamus, whereas the maternal genome favors development of the neocortex. I've suggested that some maternal-paternal conflicts can be seen within the individual between different parts of the brain favoring different sorts of actions. I don't have a good explanation of why that's occurring in the mouse; I would love to know. At a broader level, perhaps these theories have something to say about the subjective experience of internal conflicts—why we sometimes have great difficulty making up our minds. If the mind were purely a fitness-maximizing computer with a single fitness function, then the paralyzing sense of indecision we often feel would make no sense. When we are forced to make a difficult decision, it can sometimes consume all our energies for a day, even though we'd be better off making a decision one way or the other. Perhaps that can be explained as a political argument going on within the mind between different agents with different agendas. That's getting very speculative now, though.

In the future, I'd like to get back to plants. I've put a lot of work into thinking about plant life cycles, and the work I did in my Ph.D has had relatively little impact, so I'd like to go back and rethink some of those ideas. I've thought of writing a book called *Sociobotany* that would do for plants what Trivers, [E. O.] Wilson, and Dawkins did for animal behavior. Botany tends to look at the different stages in the life cycles of a plant as cooperating one with the other. But Trivers's theories of parent-offspring conflict are relevant to understanding some odd features of seed development and the embryology of plants. One of my favorite examples of this phenomenon can be seen in the seeds of pine trees and their relatives. The seed contains multiple eggs that can be fertilized by multiple pollen tubes, which are the functional equivalent of sperm. Within the seed, multiple embryos are produced that then compete to be the only one that survives in that seed. As this happens, there's intense sibling rivalry and even siblicide going on in the seed. Because of oddities of plant reproduction, the eggs that produce those embryos are all genetically identical one to the other, so all the competition among the embryos is between the genes that they get from their fathers through the pollen tube. Because of this, I expect there to be imprinting in the embryos of pine trees.

Another interesting case is found in Welwitschia, an odd plant that grows in the Namibian desert. Here, once again because of oddities of the plant's genetics, the egg cells are no longer genetically identical one to the other, and they compete with each other to produce the embryo that survives in that seed. Rather than waiting for the pollen tube to reach the eggs, the eggs grow in tubes up to meet the pollen tubes. There's actually a race to meet the pollen tubes growing down to meet the eggs. Fertilization occurs and then the embryos race back down into the seed to gain first access to the food reserves stored in the seed. This odd behavior was just a strange observation of plant embryologists, but I think the application of ideas of conflict between different genetic individuals gives a pleasing explanation of why you observe this behavior in Welwitschia but not in groups where the eggs are genetically identical to each other.

Some of these ideas also intersect with the work of evolutionary psychologists. Although I don't interact with them on a daily basis, they're keen on my work, and I follow theirs. A true psychology has got to be an evolutionary psychology. Whether every theory that goes under the name of evolutionary psychology is evolutionarily justified is a different question, but in terms of the question of whether Darwin is relevant to understanding the mind and human behavior, evolutionary psychologists have got it right. We are evolved beings and therefore our psychology will have to be understood in terms of natural selection, among other factors.

3
A Full-Force Storm With Gale Winds Blowing

Robert Trivers

[October 18, 2004]

Robert Trivers is an evolutionary biologist and sociobiologist at Rutgers University.

For the last 10 or 15 years, I've been trying to understand situations in nature in which the genes within a single individual are in disagreement—or put differently, in which genes within an individual are selected in conflicting directions. It's an enormous topic, which 20 years ago looked like a shadow on the horizon, just as about 100 years ago what later became relativity theory was just two little shadows on the horizon of physics; both blew up to become major developments. In genetics it's fair to say that about 20 years ago the cloud on the horizon was our knowledge that there were so-called selfish genetic elements in various species that propagated themselves at the expense of the larger organism. What was then just a cloud on the horizon is now a full-force storm with gale winds blowing.

An enormous amount of work is pouring out on this topic, with an appreciation that, far from being a rare exception, this is a minority phenomenon in all organisms, including ourselves, and must by logic and by evidence have been an important problem throughout the evolution of the genetic system: namely, how to control and prevent the further spread of such selfish elements. There's a dynamic relationship between these elements and the rest of the organism—or, put in genetic terms, all the other nonlinked genes in the organism, which will tend to be selected to suppress these elements and will then select on the part of the elements for further tricks in a co-evolutionary struggle.

The more we have worked on this, the more it has seemed analogous to social interactions at the level of individuals within groups. Some of the same terms we find useful in the latter context, like cooperative, selfish, spiteful, or altruistic behavior, can also be applied to genetic interactions within the individual that run into conflict with each other. There can be spiteful genes, there can be merely selfish genes, there's a whole bunch of cooperative ones, and there are narrowly altruistic ones that only help copies of themselves. This is a deep and important subject, and we are at last becoming able to see a unified whole and to relate the different parts to each other.

Genetics is extremely difficult but also very rewarding. You get an exactness out of genetics, beginning with Mendel's famous quantitative pea ratios. It's a quantitative, exact science that is beautiful, but also difficult to master. We've had a long, wonderful tradition of learning an endless series of sometimes incredible facts about the genetic systems of different organisms, without having a clear evolutionary logic for how natural selection works on the genetic system itself. This is a major avenue into that problem.

There is also a personal irony for me, in that some of those who were most vitriolic against the social theory I worked on in the 1970s were population geneticists. I expect them to be equally, or perhaps even more, displeased with this new development. Of course, that only gives me greater satisfaction.

Incidentally, the enormity of the subject required me to get a collaborator, and I was fortunate some 12 years ago to attract Austin Burt, a brilliant Canadian evolutionary geneticist, now working at Imperial College London.

I have completed the genetics work and I'm now eager to do some work in psychology. The Zeitgeist is such that we are now in the process of putting together an evolutionary psychology. As is often the case, the first hints of it are people running around saying, "Evolutionary psychology is coming! Evolutionary psychology is coming!" but they haven't actually done much work to bring it on. Now we're getting empirical work of increasingly high quality on aspects of human psychology interpreted in an evolutionary way.

The particular sub-area I'm interested in developing has to do with the structure of the mind in terms of biased information flow between the conscious and the unconscious, and the peculiar and counterintuitive fact that humans in a variety of situations misrepresent reality to the conscious mind while keeping in the unconscious either a fully accurate, or in any case more accurate, view of that which they misrepresent to the conscious mind. That seems so counterintuitive that it begs explanation. You would have thought that after natural selection ground away for 4 billion years and produced these eyeballs capable of such subtlety—color, motion detection, the details of granularity that we see—you would have perfected the organs for interpretation of reality such that they wouldn't systematically distort the information once it reached you. That seems like a strange way to design a railroad.

The function of this area of self-deception is intimately connected to deception of others. If you're trying to see through me right now, and if I'm lying about something you actually care about, what you see first, to speak loosely, is my conscious mind and its behavioral effects. You can get some sense of my mood or my affect. The quality of voice might signal stress while trying to deceive you. But it's much harder for you to figure out what my unconscious is up to. You have to make a study of my behavior, such as a spouse will do, much to your dismay at times.

One simple logic is that we hide things in our unconscious precisely to hide them better from other people, so the key interaction driving this is deception. I often talk about deceit and self-deception in the same voice, because you can't see self-deception properly if you don't appreciate its deceptive possibilities. Likewise, if you talk about deception without any reference to self-deception, then you tend unconsciously to limit yourself to consciously promoted deception, and you tend to overlook unconsciously promoted deception. Each failure to link the two topics limits one's understanding of the topic under consideration.

There is also a new area within individual deception which is related to this concept of self-deception directed toward others but which has not been worked out in a detailed way. That's the extraordinary finding that our maternal genes and our paternal genes—that is, those we inherited from our mother and those we inherited from our father—are able to conflict with each other, each acting to advance the interest of the relevant parent and his or her relatives. You can have a form of internal deception where the maternal side is over-representing maternal interests which the paternal side is discounting, and vice versa.

For some reason, I've had a deep interest in both deception and self-deception from childhood. This, of course, preceded my knowledge of evolutionary logic. In one episode I remember my mother wagging her finger in my face and saying, "Remember now, 'Judge not that ye be not judged.'" I was raised in the Presbyterian Church and, of course, this is from Matthew, who recorded Jesus saying, "Judge not that ye be not judged, for with the judgment ye pronounce shall ye be judged. And why are you worried about the mote in your neighbor's eye? First take the beam out of your own, the better to see the mote." It's an allegory for self-deception. You're so busy saying what's wrong with another person, you hypocrite. Get rid of what's wrong with you first, and don't project it onto the other person. That was a lifelong meditation my mother gave me, so there must have been something in my behavior.

The great evolutionist Ernst Mayr would say to me, "It's very appropriate you're interested in self-deception, since you sure practice a lot of it." At first I didn't know what to say, and then came up with the notion that it's exactly the people who struggle with their self-deception who you'd expect to find the problem interesting and maybe make some progress on it. Those unafflicted by it might have low insight and low motivation.

I also remember from my childhood that there was a prized item in a store that had a lot of toys for children. I think it was a knife, but I know it cost $6. I saved up $6 plus the 2 cents for tax back then, which took me a while under my father's regime of reimbursement for yard work. I went into the store and gave them the $6.02, and the man behind the counter said it cost $7.

I said, "What are you talking about?" and he said, "It's seven dollars. It says so on the sign in the window." I said, "Nonsense, the sign says six dollars!"

We went outside and he showed me the sign, and it said $6.98, with the .98 written in small letters. I remember arguing with him, asking what sense it made to misrepresent the cost of this by two pennies, so that you had to do this extra addition. He said it was very common. I remember walking around in a daze for weeks, looking at signs and thinking about the amount of arithmetic this was generating. A lot of times you had to add the two pennies, because it actually mattered, as in my own sad case. I don't know if it changed my life, but I know I had an early intense consciousness about the costs of deception and also about the importance of self-deception.

When I came to Harvard, I ended up in U.S. history, after beginning as a mathematics major. I left in despair and disgrace and was going to be a lawyer, so you studied U.S. history. This was the early 1960s, while the Vietnam disaster was starting to take shape, and was produced by Harvard people in JFK's cabinet. We were reading books like *America, Genius for Democracy,* or something like that. You didn't even have to read the book, since the title had the content. All of this U.S. history was really self-glorification. I couldn't imagine devoting your life to this kind of enterprise.

I had a breakdown as a junior, so I finally took a psychology course as a senior. I couldn't believe that these people were pretending to have a science, when all they had was a series of competing guesses about how human beings developed. There were learning psychologists, depth or Freudian psychologists, and social psychologists. It wasn't a unified discipline, it had no unifying paradigm, and it was not hooked up to an underlying science—i.e., biology—just as biology sits on chemistry, which sits on physics. I thought it was hopeless and went away. When I learned evolutionary logic and then animal behavior, which I was assigned to learn and then render for children, I realized that the basis for psychology is evolutionary logic. The value of animal behavior is that you cast your net more widely.

My mental breakdown prevented me from getting into Yale for law school and created problems with my fallback position at the University of Virginia. I decided not to go to there, because I didn't like how they were handling the medical records they insisted they had to have. I happened, instead, to get a job writing and illustrating children's books for the new social sciences, which followed the new math and the new physics, which followed *Sputnik* in 1957. That was our attempt to catch up with the Soviets in science, and then in social science. The reason you probably never heard of it was because the whole course was killed by a set of Southern congressmen because it was alleged that we taught (1) sex education (we had pictures of animals copulating), (2) evolution by natural selection as fact (which was true), and cultural relativity (i.e., respect for other cultures). The whole thing was killed, but it introduced to me to evolutionary logic.

It was a marvelous company and a beautiful setup, in retrospect. The company allowed you to do a lot of reading right there in the office to learn what you needed to know. In my case, they assigned a biologist named Bill Drury from the Massachusetts Audubon Society who assigned papers in the library on a given topic and then critiqued them for me. He was paid $75 an hour in 1966, which is at least $200 or $300 an hour now. For an academic or the head of the Massachusetts Audubon Society, that's some sweet money, so I had the ideal situation where you could consume two hours of your teacher's time with no guilt at all. So I had a private tutor in biology, paid for by my employer for two years.

He took me to see Ernst Mayr to try to talk me into being a graduate student. I came out of mathematics, where if you haven't done any math by the time you're twenty-three, it's unlikely that you will be a mathematician. I thought that to be a biologist I should have been studying insects from the time I was four, but Drury would just say that whenever you ask a biologist an interesting question, he won't know the answer.

When Drury took me to see Mayr, I liked right away that

Ernst had a small office off the bigger office of his secretary. He had his own private office elsewhere, but as head of the Museum of Comparative Zoology he occupied a little space and she occupied the bigger space. He told me about Dick Estes, who at thirty-eight had gone back to school in biology and had just finished a good thesis on the wildebeest. He was very encouraging. Then there was a funny moment when he asked, "Who do you want to work with?"

I didn't know anything, so I said, "Konrad Lorenz." He read my personality right and said, "He's too authoritarian for you. That isn't gonna work. Who else?"

I said, "What about Niko Tinbergen?" And he said, "He's only repeating now in the '60s what he already showed in the '50s."

I'm a relatively quick learner, so I said, "Well, what do you suggest, Professor Mayr?" I'll never forget his hands, going in a wide circle, as he said, "What about Harvard?"

And what about Harvard, indeed? They had a marvelous museum with all these fossils and pinned insects. They didn't have any animal behaviorists, but my teacher, Drury, who was a behaviorist, convinced me that that was even an advantage. He said, "Why would you want to take a course on field methodology and learn how to put a band on a bird, which you can learn better in the field from a teacher. What you want to learn is evolutionary biology." Irv DeVore would have allowed me to come straight in as a graduate student in anthropology. If I had done that, I wouldn't have had to borrow a bunch of money and take a year of courses in biology, but I knew that would have been a shortsighted decision. I knew all the ideas and the power were coming out of biology, so that's what I should learn.

I had had no chemistry either, so Ernst suggested I take chemistry at night school at Boston University, since it was too hard at Harvard. I did everything Ernst told me, so although I was working this job, at five o'clock I'd bicycle to Boston University; I took one semester of chemistry. In the second semester, I had an opportunity to finish the course or to watch caribou in the Arctic for a month. I thought that such a trip was more valuable to my long-term development than second-semester chemistry. Incidentally, I avoided chemistry entirely. I'm one of the few Ph.Ds in this country who's never had a course in organic chemistry, because you cannot get a bachelor's in this country without an organic degree. I don't have a bachelor's.

When I came to Harvard, you had to take a whole series of mini-tests. There were 16 of them—physics, math, botany, chemistry, etc. Naturally I failed. When the prescription committee met, it was stacked with evolutionists and Ernst was the head. He insisted that the committee not only prescribe me organic chemistry—that is, that I couldn't get my Ph.D unless I passed that with a B or better—but also prescribed me a knowledge of biochemistry, because that's the payoff for organic chemistry. The argument got hot, and I tried to intercede at one moment, until I realized, "This has nothing to do with you. Shut up and sit back." Finally Mayr says, "By God, I agree with you! We should not prescribe organic unless we prescribe biochem, and since we will not prescribe biochem we will also not prescribe organic." He put it up to a vote and it came back 5 to 2. It was as if the heavens had opened and the Lord himself smiled at me and said, "You are my chosen one." Then Mayr leaned across the table and said, "But Bob, we strongly urge you to take organic chemistry." And I said, "Professor Mayr, I'm already signed up for the course," which I was. I sold the book that afternoon and burned the little Tinker toys you had to buy.

I came to Harvard originally as a special student who had never had biology. That fall I was taking a course in cell biology, a course in invertebrate biology, and a botany course. I used to sit in my bed at night with these biology books and a dictionary trying to figure it out—and I used to have "word salad" dreams for a couple of months—"the cnidoblast of the intestinal cells of the apple's pollen" and so on. After two or three months, the subjects all separated cleanly into their separate sub-areas, and it wasn't so bad.

The guy who really got me focused properly was Richard Lewontin, a geneticist who hated my work, helped make sure that Harvard didn't give me tenure right away when I wanted it, and will undoubtedly hate my genetics work. Dick Lewontin came to Harvard when I was a first-year graduate student in 1969 to give a talk on the new methodology of isozyme work that John Hubby had worked out. That was the first DNA technology that would allow you to do paternity analysis. It was not so much nailing down who your real father was that was exciting to us biologists, but the fact that you could quantify the degree of genetic variation in nature. I was introduced to the guy ahead of time by Ed [E. O.] Wilson, and he [Lewontin] dumped all over me because of a nasty paper I'd written about some mathematical ecologists. I took an instant dislike to him—he had a rather arrogant style—and I was hoping he would fall flat on his face. He was introduced by Mayr and gave a superb talk. My joke is that at the end he flipped his chalk 30 feet in the air and caught it in his breast pocket. He didn't really do that, but he might as well have. Everything else was in place, including intellectual content and showmanship.

Halfway through the talk I was feeling an intense internal pain, because although I disliked Lewontin he was doing a great job. I did some quick thinking and realized there was no future in this negative paper I'd written on mathematical ecology. I had no positive thoughts on the subject, nor did I have the talents that would make that area pay off in my life, so I decided to do no more work on it. I was tempted to write it up, because the Harvard professors wanted me to nail some people they disliked, but I decided not to waste any more time on it. It was clear to me during Lewontin's talk that if the work was as bad as I said it was, then my critique, if it was published, would disappear from sight along with their work, like a barnacle on a whale.

The only value comes if you have something positive to do, and it's important to match both your own interests and abilities to what you decide to work on. I often wonder how many scientists end up spending 5 or 10 years in mathematical ecology because of some accident of a paper they wrote. They never quite see that they've been running around in the woods for 30 years and have no intuitions of any use about the way nature is set up. I was also too lazy to learn any new mathematics, but I was blessed with a certain degree of psychological and social insight and had been interested in these issues from an early age. That's analogous to running around in the woods for 30 years, if you're going to sit down and write social theory.

I asked myself, "What ideas do you have that are worth developing?" I started thinking about the obvious concept, "If you scratch my back, I'll scratch yours," and began to wonder about how to make reciprocal altruism work in an evolutionary way, stating the argument in a form that didn't limit it to humans. That was worth throwing some time into.

Bill Drury was an ornithologist and had told me to watch pigeons. From watching them every night, I knew they had a double standard. During the day, males obsessed about the chance that another male might get in there with his female. But at the same time, that male was hustling other females whenever he got a chance. Out of this came a general theory for the evolution of sex differences, parental investment in sexual selection. The paper that developed out of this has been cited more than 4,000 times, because much work on sex differences—on role, style, and all the behavioral stuff—refers back to that original paper, especially if they cite the concept of parental investment. It actually appears to be cited more often per month now than earlier in its life.

People have asked me about the connection between deceit and self-deception in evolutionary biology. Certainly I was conscious of deception right away, and my teacher Bill Drury was very helpful. I might easily have had an inferior teacher who was unconscious of the degree of deception in other animals. Then I might have made the mistake some biologists did of talking as if we were the only deceiving creatures and that therefore this trait had to do with language. I knew from early on that it was much more general, and probably I learned that from Bill.

People have pointed out to me that in my first paper on reciprocal altruism, there's argumentation that refers to keeping the feeling unconscious so as not to have it detected. My first consciousness of this occurred when I went on a field trip to India and Africa with Irv DeVore, Harvard's celebrated baboon man. I had a brainstorm for a couple of weeks, and ended up in a hospital for 10 days afterward. I thought about hardly anything other than parent-offspring conflict and deceit and self-deception. Some of it, in retrospect, was fruitless efforts to map Freud onto parent-offspring conflict and deceit and self-deception, whose deeper ramifications I was just appreciating. In later years I came to think it was worthless, that it was better to start without Freud and certainly not go through the genuflections that Freudians seem addicted to.

The joke on me is that I never developed it. I was supposed to give a paper on the logic of self-deception in 1978 at a conference put on by the Royal Society in London. I wrote an abstract, which was published. I came across it several years ago and said, "I would like to read that paper." Back then I was young and strong and could write an abstract eight months before I was due to give a talk and just plan to fill in the blanks, but I haven't worked out the details of some of those assertions in that abstract yet. It's just floating around there. I never gave the paper, partly because my wife was about to give birth to twins, but more particularly because the Royal Society would only fly you to England. I guess the assumption is that once you got there, why would you want to leave? I was very sensitive to financial exploitation of academics, which is rife. It certainly was at Harvard, where we were grossly underpaid as junior faculty, so I tended never to do something whose financial arrangements I didn't like.

I never went there, never gave the talk, never developed the paper, and that's a great shame, because one of the virtues of thinking a topic through to some degree of development is that you will generate a literature which will come back and illuminate the topic for you. Even if you're thinking in purely self-interested terms and write a paper on reciprocal altruism, there's a huge literature now on the subject. Only part of it is generated from that paper, but still a good part was generated from that paper, and I learned back from it. I often think of the paper not written and the literature that did not develop, especially as I sat down at the end of the '90s and wrote a quickie paper on self-deception to try to bring the field up to date. I was staggered at how little progress had been made since the last time I looked at the subject, 20 years before. That was the cost of never writing a paper. It didn't need to be as elemental or as important as some of the other ones, but just writing it to put it on other people's laps would have generated a response.

I was very fortunate. Lewontin once referred to me to a bunch of graduate students as an intellectual opportunist. He meant that to be negative, but I laughed. What else makes sense in this short life? I *was* an intellectual opportunist. All of these social topics remain undeveloped because of this species advantage "paradigm" that had lain over the field like a human blanket. Reciprocal altruism, parental investment, sexual selection, sex ratio of offspring, parent-offspring conflict were all topics sitting there waiting to be developed. I just grabbed the chance.

But I had to get away from Harvard. I was a graduate student at Harvard from 1968 to 1972, then I started teaching at Harvard in 1973, until 1978. I was not denied tenure, I just needed more money for what I was doing, which was lecturing to 530 students, with 12 graduate students as teaching assistants, and so on. Most of the junior faculty at Harvard taught 15 to 20 students in an undergraduate class and 4 in a graduate class. They were being eaten alive. Harvard wasn't paying me enough to replenish what was being taken out of me biologically. Never mind not being paid enough to have any reproductive success of my own.

There were two factors at work. First, no one knew about my work, which was really nice. It's good to be flying underneath everyone's radar and then publish a paper you know is important. But with Ed Wilson's *Sociobiology*, which embraced some of my ideas, the proverbial you-know-what hit the fan, and there was a political stink. People were very upset then. The Vietnam War was still going on, for God's sake, in 1975, and people were very politically conscious and pseudo-conscious. So I became well known. That was ego-distorting. And then I was at Harvard, which is a separate kind of ego-distorting environment, just because Harvard professors can't help being full of themselves.

I had been in Cambridge 17 years, 15 of them at Harvard with 2 years off for good behavior—that was between the undergraduate years and the going back as a special student. So I had to leave Harvard if only to air out my psyche, but I did *not* have to pick UC Santa Cruz, perhaps the second worst school in its class in the country. Lord, what a place! It was a very, very bad fit for me, and a dreadful 16 years. Thank God I came back East.

I intend to throw myself full time into deceit and self-deception now. It's a topic I've been conscious of in an evolutionary way for at least 30 years. I've published bits and pieces throughout the years. I used to get up and joke that I was embarrassed lecturing on it because I'd been practicing it for the last 30 years instead of thinking about it. I don't like to make that joke any more. I want to get completely on top of the subject, and I want to do a major piece on it. And I don't want it to be just for an academic crowd, because the topic is everywhere. It's in every human being's life, and anybody who's half conscious is aware of it in others and themselves. One cannot read the newspaper without being conscious of the importance of deceit and self-deception in national and international affairs.

I have the free time now to do whatever I want, flying underneath the radar. I work much better in a much humbler posture, where people basically don't know who I am, or if they do it doesn't mean any particular thing to them. I'm not invited a lot of places. Otherwise I would have my time eaten up, but I'm not in that situation. I can throw myself with full energy into it, and I plan to.

I'm particularly excited by the fact that almost every month neurophysiology is coming out with a result of direct relevance to the topic of self-deception. The psychologists have invented skillful new techniques at getting at pre-conscious or unconscious processes that are exciting. There is an empirical scientific world building up now that did not exist 20 years ago and that can constrain and guide our thinking.

I sometimes contrast this topic to genetics. Genetics, as I mentioned, is intrinsically difficult, but it's exact. If you take the energy and the time to master it, you will get rewards for it. You will actually know real things and be able to point to it and know what you're talking about. Deceit and self-deception, by their nature, are topics that tend to be hidden from view. They are difficult to pinpoint, even as to how you define them. It's a different kind of intellectual problem than genetics. It's much easier to master what's known, because in scientific terms not an awful lot *is* known. But if you have to think carefully in terms of the logical distinctions you make, there's now an emerging body of empirical data that, as I say, can constrain your thinking and guide it. If you're not constrained, the topic is too big and the possibilities are too great. You have to be able to say, "No, we're going to exclude this half, or this two-fifths, of reality because of this result, and we think this is where things are important, because the data point in that direction."

The time is ripe, although it will be riper 5 to 10 years from now. Academics are always saying this is the perfect study area for this, the perfect species to do that, or the perfect time for this book. Well, this isn't the perfect time, but at the very least it will point people toward relevant empirical work. It's a function of how much actual thinking you do. Like everything else, this is not a topic where seat-of-the-pants thinking, or a few polished anecdotes no matter how amusing, are going to carry the day. The topic has got to be thought through, carefully and systematically, and I am ripe to do it.

4
What Evolution Is

Ernst Mayr

[October 31, 2001]

Ernst Mayr (1904–2005),was a leading evolutionary biologist, taxonomist, and ornithologist.

Introduction by Jared Diamond [excerpted from his Introduction to Mayr's *What Evolution Is* (New York: Basic Books, 2001)]

When the first bird survey of the Cyclops Mountains was carried out I found it hard to imagine how anyone could have survived the difficulties of that first survey of 1928, considering the already-severe difficulties of my second survey in 1990. That 1928 survey was carried out by the then-23-year-old Ernst Mayr, who had just pulled off the remarkable achievement of completing his Ph.D thesis in zoology while simultaneously completing his pre-clinical studies at medical school. Like Darwin, Ernst had been passionately devoted to outdoor natural history as a boy, and he had thereby come to the attention of Erwin Stresemann, a famous ornithologist at Berlin's Zoological Museum. In 1928 Stresemann, together with ornithologists at the American Museum of Natural History in New York and at Lord Rothschild's Museum near London, came up with a bold scheme to "clean up" the outstanding remaining ornithological mysteries of New Guinea, by tracking down all of the perplexing birds of paradise known only from specimens collected by natives and not yet traced to their home grounds by European collectors. Ernst, who had never been outside Europe, was the person selected for this daunting research program.

Ernst's "clean-up" consisted of thorough bird surveys of New Guinea's five most important north coastal mountains, a task whose difficulties are impossible to conceive today in these days when bird explorers and their field assistants are at least not at acute risk of being ambushed by the natives. Ernst managed to befriend the local tribes, was officially but incorrectly reported to have been killed by them, survived severe attacks of malaria and dengue and dysentery and other tropical diseases plus a forced descent down a waterfall and a near-drowning in an overturned canoe, succeeded in reaching the summits of all five mountains, and amassed large collections of birds with many new species and subspecies. Despite the thoroughness of his collections, they proved to contain not a single one of the mysterious "missing" birds of paradise. That astonishing negative discovery provided Stresemann with the decisive clue to the mystery's solution: All of those missing birds were hybrids between known species of birds of paradise, hence their rarity.

From New Guinea, Ernst went on to the Solomon Islands in the Southwest Pacific, where as a member of the Whitney South Sea Expedition he participated in bird surveys of several islands. . . . A telegram then invited him to come in 1930 to the American Museum of Natural History in New York to identify the tens of thousands of bird specimens collected by the Whitney Expedition on dozens of Pacific Islands. . . . Ernst Mayr's "explorations" of bird specimens in museums were as important as his fieldwork in New Guinea and the Solomons in forming his own insights into geographic variation and evolution. In 1953 Ernst moved from New York to Harvard University's Museum of Comparative Zoology. . . . For scholars studying evolution and the history and philosophy of biology, Ernst's hundreds of technical articles and dozens of technical books have been for a long time the standard reference works.

But in addition to gaining insights from his own fieldwork in the Pacific and from his own studies of museum bird specimens, Ernst has collaborated with many other scientists to extract insights from other species, ranging from flies and flowering plants to snails and people. One of those collaborations transformed my own life, just as the meeting with Erwin Stresemann transformed Ernst's life. While I was a teenaged schoolboy, my father, a physician studying human blood groups, collaborated with Ernst in the first study proving that human blood groups evolve subject to natural selection. I thereby met Ernst at dinner at my parents' house, was later instructed by him in the identification of Pacific island birds, began in 1964 the first of 19 ornithological expeditions of my own to New Guinea and the Solomons, and in 1971 began to collaborate with Ernst on a massive book about Solomon and Bismarck birds that we completed only this year, after 30 years of work. My career, like that of so many other scientists today, thus exemplifies how Ernst Mayr has shaped the lives of 20th-century scientists: through his ideas, his writings, his collaborations, his example, his lifelong warm friendships, and his encouragement.

EDGE: To what extent has the study of evolutionary biology been the study of ideas about evolutionary biology? Is evolutionary biology the evolution of ideas, or is it a fact?

ERNST MAYR: That's a very good question. Because of the historically entrenched resistance to the thought of evolution, documented by modern-day creationism, evolutionists have been forced into defending evolution and trying to prove that it is a fact and not a theory. Certainly the explanation of evolution and the search for its underlying ideas has been somewhat neglected, and my new book, the title of which is *What Evolution Is*, is precisely attempting to rectify that situation. It attempts to explain evolution. As I say in the first section of the book, I don't need to prove it again: Evolution is so clearly a fact that you need to be committed to something like a belief in the supernatural if you are at all in disagreement with evolution. It is a fact, and we don't need to prove it anymore. Nonetheless we must explain why it happened and how it happens.

One of the surprising things I discovered in my work on the philosophy of biology is that when it comes to the physical sciences, any new theory is based on a natural law. Yet as several leading philosophers have stated, and I agree with them, there are no laws in biology like those of physics. Biologists often use the word "law", but for something to be a law, it has to have no exceptions. A law must be beyond space and time, and therefore it cannot be specific. Every general truth in biology, though, is specific. Biological "laws" are restricted to certain parts of the living world or certain localized situations, and they are restricted in time. So we can say that there are no laws in biology, except in functional biology, which, as I claim, is much closer to the physical sciences than the historical science of evolution.

EDGE: Let's call this Mayr's Law.

MAYR: Well, in that case, I've produced a number of them. Anyhow, the question is, If scientific theories are based on laws and there aren't any laws in biology, well, then, how can you say you have theories, and how do you know that your theories are any good? That's a perfectly legitimate question. Of course, our theories are based on something solid, which are concepts. If you go through the theories of evolutionary biology, you find they're all based on concepts such as natural selection, competition, the struggle for existence, female choice, male dominance, etc. There are hundreds of such concepts. In fact, ecology consists almost entirely of such basic concepts. Once again you can ask, How do you know they're true? The answer is that you can know this only provisionally, by continuous testing, and you have to go back to historical narratives and other non-physicalist methods to determine whether your concept and the consequences that arise from it can be confirmed.

EDGE: Is biology a narrative based on our times and how we look at the world?

MAYR: It depends entirely on when in the given age of the intellectual world you ask these questions. For instance, when Darwin published *On the Origin of Species*, the leading Cambridge University geologist was [Adam] Sedgwick, and Sedgwick wrote a critique of Darwin's *Origin* that asked how Darwin could be so unscientific as to use chance in some of his arguments when everyone knew that God controlled the world. Now, who was more scientific, Darwin or Sedgwick? This was in 1860, and now, 140 years later, we recognize how much this critique was colored by the beliefs of that time. The choice of historical narratives is also time-bound. Once you recognize this, you cease to question their usefulness. There are a number of such narratives that are as ordinary as proverbs and yet still work.

EDGE: Darwin is bigger than ever. Why?

MAYR: One of my themes is that Darwin changed the foundations of Western thought. He challenged certain ideas that had been accepted by everyone, and we now agree that he was right and his contemporaries were wrong. Let me just illuminate some of them. One such idea goes back to Plato, who claimed there were a limited number of classes of objects and each class of objects had a fixed definition. Any variation between entities in the same class was only accidental, and the reality was an underlying realm of absolutes.

EDGE: How does that pertain to Darwin?

MAYR: Well, Darwin showed that such essentialist typology was wrong. Darwin, though he didn't realize it at the time, invented the concept of biopopulation, which is the idea that the living organisms in any assemblage are populations in which every individual is uniquely different, which is the exact opposite of such a typological concept as racism. Darwin applied this populational idea quite consistently in the discovery of new adaptations, though not when explaining the origin of new species.

Another idea Darwin refuted was that of teleology, which goes back to Aristotle. During Darwin's lifetime, the concept of teleology, or the use of ultimate purpose as a means of explaining natural phenomena, was prevalent. In his *Critique of Pure Reason,* Kant based his philosophy on Newton's laws. When he tried the same approach in a philosophy of living nature, he was totally unsuccessful. Newtonian laws didn't help him explain biological phenomena. So he invoked Aristotle's final cause in his *Critique of Judgment*. However, explaining evolution and biological phenomena with the idea of teleology was a total failure.

To make a long story short, Darwin showed very clearly that

you don't need Aristotle's teleology, because natural selection applied to bio-populations of unique phenomena can explain all the puzzling phenomena for which previously the mysterious process of teleology had been invoked. The late philosopher Willard Van Orman Quine, who was for many years probably America's most distinguished philosopher, told me about a year before his death that as far as he was concerned, Darwin's greatest achievement was that he showed that Aristotle's idea of teleology, the so-called fourth cause, does not exist.

EDGE: Is this an example of Occam's razor?

MAYR: It's that in part as well, but what's crucial is the fact that something that can be carefully analyzed, like natural selection, can give you answers without your having to invoke something you cannot analyze, like a teleological force.

Now, a third one of Darwin's great contributions was that he replaced theological, or supernatural, science with secular science. [Pierre-Simon] Laplace, of course, had already done this some 50 years earlier when he explained the whole world to Napoleon. After his explanation, Napoleon asked, "Where is God in your theory?" And Laplace answered, "I don't need that hypothesis." Darwin's explanation that all things have a natural cause made the belief in a creatively superior mind quite unnecessary. He created a secular world, more so than anyone before him. Certainly many forces were verging in that same direction, but Darwin's work was the crashing arrival of this idea, and from that point on, the secular viewpoint of the world became virtually universal.

So Darwin had an amazing impact, not just on evolutionary theory but on many aspects of everyday human thought. My firm belief is that each period in world history has a particular set of ideas that are the Zeitgeist of that period. And what causes this Zeitgeist? The answer usually is that there are a couple of important books that have been responsible for everybody's thinking. The number-one book in this realm is, of course, the Bible. Then for many years the answer might have been Karl Marx's *Das Kapital*. There was a short period when Freud was mentioned—though I don't think he's mentioned anymore by anyone besides the Freudians. The next one—and there is no doubt in my mind that Darwin's *Origin of Species* was the next one—not only secularized science, gave us the story of evolution, but also produced hosts of basic theoretical concepts, like bio-populationism and, as I showed, the repudiation of teleology. No one before Darwin had introduced those ideas or advanced them so forcefully.

EDGE: Not even the scientific community outside of evolutionary biologists?

MAYR: No. They weren't brought up with these ideas, though scientists like T. H. Huxley probably felt, as he said, "How stupid of me not to have thought of it."

EDGE: How do you account for the fact that in this country, despite the effect of Darwinism on many people in the scientific community, more and more people are god-fearing and believe in the six days of creation?

MAYR: You know you cannot give a polite answer to that question.

EDGE: In this venue we appreciate impolite, impolitical answers.

MAYR: They recently tested a group of schoolgirls. They asked, "Where is Mexico?" Do you know that most of the kids had no idea where Mexico is? I'm using this only to illustrate the fact that—and pardon me for saying so—the average American is amazingly ignorant about just about everything. If he were better informed, how could he reject evolution? If you don't accept evolution, then most of the facts of biology don't make sense. I can't explain how an entire nation can be so ignorant, but there it is.

EDGE: I understand there's a facsimile of the first (1859) edition of Darwin's *Origin of Species*.

MAYR: Yes, and this is an interesting story. Darwin's importance has only been gradually acknowledged. Even 50 years ago, Darwin was just one of those names you learned was kind of important. That was it. Nobody read him. Well, I published a very successful book for Harvard University Press in 1963 and this gave me the courage to go to the director of Harvard Press, Tom Wilson, and say to him, "Tom, I have a great wish, a heart's desire, and that is to see a facsimile edition of the first edition of the *Origin of Species*. We have facsimile editions of all the great classics, but we haven't got one for Darwin." So he said, "All right, all right, we'll do it for you, even though we'll probably lose money, since who's going to buy it?" In 1964 they published this facsimile edition. And at that time, the first few years, I guess they sold about a couple hundred a year, but much to everybody's surprise, sales did not drop off after all the libraries had their facsimile edition; rather, they picked up. After a while, they sold over 1,000 a year, and then about six or seven years ago I was informed by Harvard Press that they had for the first time sold 2,500 copies. The last two years I have a report that they sold

3,000 copies a year! This shows you how an interest in Darwin has been steadily growing in spite of the great majority of ignorant people. People are beginning to want to know what Darwin really said, which for me is an absolutely marvelous development. You know, there's an interesting side note that, as a publisher, you might be interested in. In the first edition of the *Origin of Species* there's not a single misprint. What a document of the workmanship in 1859!

EDGE: Where do you think Darwinism is going to go in the next 50 years?

MAYR: Well, Darwinism will not have to do any going, because it's already here. In the last 50 years, ever since the evolutionary synthesis of the 1940s, the basic theory of Darwinism has not changed, with perhaps one exception—that is the question of the target of selection. What's the object of a selective act? For Darwin, who didn't know any better, it was the individual—and it turns out he was right.

An individual either survives or doesn't, an individual either reproduces or doesn't, an individual either reproduces successfully or doesn't. The idea that a few people have about the gene being the target of selection is completely impractical; a gene is never visible to natural selection, and in the genotype it is always in the context with other genes, and the interaction with those other genes makes a particular gene either more favorable or less favorable. In fact, [Theodosius] Dobzhansky, for instance, worked quite a bit on so-called lethal chromosomes, which are highly successful in one combination and lethal in another. Therefore people like [Richard] Dawkins in England who still think the gene is the target of selection are evidently wrong. In the 30's and

40's, it was widely accepted that genes were the target of selection, because that was the only way they could be made accessible to mathematics, but now we know that it's the whole genotype of the individual, not the gene. Except for that slight revision, the basic Darwinian theory hasn't changed in the last 50 years.

EDGE: Where does the generation of William Hamilton, George Williams, and John Maynard Smith fit in?

MAYR: Hamilton never denied the primacy of the individual. In the case of G. C. Williams, I have come to the unhappy conclusion that not many of the proposals of his best known book, *Adaptation and Natural Selection* (1966) are valid.

EDGE: All right, so Darwinism isn't going to change in 50 years, but the people writing about it certainly are changing.

MAYR: Every year one if not two books come out about Darwinian theory. Many of them are favorable, which is fine; yet many others attempt to improve or revise Darwin's original ideas, coming up with some so-called new theory that is invariably total nonsense.

EDGE: I can imagine what you think about evolutionary psychology.

MAYR: Not necessarily! To tell the truth, I don't know much about it, but I've heard there's a field called evolutionary epistemology. They use a simple Darwinian formula that can be stated in a single sentence. If you have a lot of variation, more than you can cope with, only the most successful will remain. That is how things happen. In epistemology and countless other fields. Variation and elimination.

EDGE: Who's notable in that field?

MAYR: Quite a few people, though I can't recall their names right now. Suffice it to say that there are many more evolutionary epistemologists in Germany and Austria than in this country.

EDGE: It seems to me that Darwin is much better known in England than in the United States. Books about Darwin sell well, and people debate the subjects. Here in America what passes for intellectual life doesn't necessarily include reading and having an appreciation of Darwin.

MAYR: Yet the funny thing is, if in England you ask a man in the street who the greatest living Darwinian is, he will say Richard Dawkins. And indeed, Dawkins has done a marvelous job of popularizing Darwinism. But Dawkins's basic theory of the gene being the object of evolution is totally non-Darwinian. I would not call him the greatest Darwinian. Not even Maynard Smith. Maynard Smith was raised in math and physics, and he was an airplane engineer in the last war. For the most part, he still thinks like a mathematician and engineer. His most successful contribution to evolutionary biology has been applying so-called game theory to evolution. Personally I have—and now I perhaps expose myself to a great deal of criticism, but regardless—I have always been a little unhappy about that application of game theory. What animal ever, in a confrontation, would say, "Now, let me figure it out—Would it be better to be timid or would it be better to be bold?" That's not the way organisms think. You get—and somebody would have to work this out, since I'm not a mathematician—exactly the same result if you have a population with every animal acting with a different mixture of timidity and boldness. Individuals at one end of the curve are very timid and have little boldness, individuals in the middle of the curve have an appropriate mixture of timidity and boldness, and individuals at the other end of the curve are very bold. Somewhere in between, in a given environment with a given set of enemies and competitors, is the best mixture of the two tendencies. You get the same results with game theory, but in my opinion the better solution has a more biological, Darwinian approach.

EDGE: How can the evolution of human ethics be reconciled with Darwinism? Doesn't natural selection always favor selfishness?

MAYR: If the individual were the only target of selection, this would indeed be an inevitable conclusion. However, small social groups that compete with each other, such as the groups of hunter-gatherers in our human ancestry, were, as groups, also targets of selection. Groups whose members actively cooperated with each other and showed much reciprocal helpfulness had a higher chance for survival than groups that did not benefit from such cooperation and altruism. Any genetic tendency for altruism would therefore be selected in a species consisting of social groups. In a social group, altruism may add to fitness. The founders of religions and philosophies erected their ethical system on this basis.

EDGE: What important questions have I not asked you?

MAYR: One question that is a difficult one to answer is whether the Darwinian framework is robust enough to remain the same for many years, which I think it is, yes. The real question is what the burning issues in evolutionary biology are today. To answer that, you've got to get back into functional biology. Take, for instance, a particular gene. Say this gene makes amino acids that determine which side of the egg is to become the anterior end of the larva and which will become the rear. We know that's what it does, but how it can do that is something about which we don't have the slightest clue. That's one of the big problems, but it's in the realm of proteins and functional biology rather than of DNA and evolutionary biology.

In evolutionary biology we have species like horseshoe crabs. The horseshoe crab goes back in the fossil record over 200 million years without any major changes. So obviously they have an invariant genome type, right? Wrong, they don't. Study the genotype of a series of horseshoe crabs and you'll find there's a great deal of genetic variation. How come, in spite of all this genetic variation, they haven't changed at all in over 200 million years, while other members of the ecosystem in which they were living 200 million years ago are either extinct or have developed into something totally different? Why did the horseshoe crabs not change? That's the kind of question that stumps us at present.

Then there are issues that no one besides a few biologists can fully fathom. Like how and why do prokaryotes, bacteria that have no nucleus, differ in their evolution from eukaryotes, organisms that do have a nucleus. Eukaryotes have sexual reproduction, genetic recombination, and well-formed chromosomes, whereas prokaryotes have none of the above. So how do they get genetic variation, which they must have in order to survive, according to the principle of natural selection? The answer is that prokaryotes exchange genes with each other unilaterally; one bacterium injects a set of DNA into another bacterium, which is an amazing process. Genes of course also go from one chromosome to another via this old-fashioned process that all bacteria use to reproduce. Beyond that, we don't really know how much such gene transfer occurs in higher organisms.

EDGE: A number of years ago, I was talking to a German publisher about a new book on Darwinism. "I can't publish it," he said. "It's just too hot to handle." Why is Darwin so dangerous, to use Dan Dennett's phrase?

MAYR: I have a good deal of contact with some very good young German evolutionary biologists, and I'm constantly amazed at how preoccupied they are with political concerns. It's just that they have gone through a series of political changes, from the Weimar Republic to the Nazi period, Soviet occupation, the DDR, and finally a united Germany, and throughout this time everything has always been colored by politics. People got their jobs because they were Nazis, or because they were anti-Nazis, and so forth. They have to find a way to purge this from their system. In Germany, they scrutinize all leaders in a field and check all the records as to whether they had been Nazis, which Nazi organizations they might have belonged to, whether they published either papers or books indicating that they had been Nazis or Communists, etc.

They think they have to do all this cleansing of science so that people can't say, "Well, you didn't tell us that so-and-so was a Nazi," or a Communist. Scientists have to cope with that. On the other hand, translations of my books published in Germany have been very successful. In fact, one of them is so successful that the

German printing has run out and I can't persuade the publisher to republish it. He asks why he should publish another German edition of the book when everybody reads the English edition. Which is true.

5
Genetics Plus Time

Steve Jones

[March 26, 2000]

Steve Jones is Emeritus Professor of Human Genetics, University College London.

STEVE JONES: The small questions I'm asking myself have to do with genetics of snails and fruit flies. Not, I suppose, of much general interest. However, they're a sub-set of a larger question, which is, "Is life simple?" And the answer is probably simpler than you would have imagined, because the rules of evolution are straightforward, and most attempts to bend or modify them have in the end turned out to be fairly unnecessary. It does look as if Darwin was, more or less, right. Most new discoveries fit well into his ideas. At the end of the century, biology looks like a more straightforward science than it did even 20 years ago, which I find a bit surprising—because, to the public, life seems fundamentally a mess. Of course, if you concentrate only on the details, they get more and more complicated. The DNA sequence is more of a mess than anyone would have ever imagined; it's not a pretty sight. But descent with modification, as Darwin put it—or genetics plus time, as we can rephrase him today—is still the foundation of life. Biology is not like physics; Newtonian physics is, in a deep sense, wrong, whereas Mendelism and Darwinism are in a deep sense right.

EDGE: What are you trying to do to persuade the public of that?

JONES: I've had the rather daring—some people might say arrogant—idea of rewriting the *Origin of Species* itself, in my new book, *Darwin's Ghost.* The idea was to take what Darwin called his "long argument" and reconstitute it with the facts of 1999 rather than 1859. As I say to annoy my publishers, it may be a rotten book, but it's a great idea—and I was amazed how well Darwin's argument stood up. We're thinking of a special pitch-impregnated edition for sale in Kansas so that it burns well, but it would be nice to think that a few creationists might read it before they condemn it to the flames. However, the difficulty with arguing with anti-rationalists is that they're not susceptible to rational argument. People like Stephen Gould have done a noble job in trying to apply rational argument. But most of them really will not be persuaded by whatever facts you present them with, so perhaps my book will have zero effect on them.

The odd thing about the Kansas fuss—and the whole creationist movement in the States—is how new it is. People always assume that when the *Origin* was published the streets ran with blood, city blocks burst into fire, churches collapsed, and hundreds hanged themselves in despair.

Of course, that wasn't true at all. There was some earnest debate among intelligent people, and by the end of the 19th century most religious people, both here and in the States, had managed to come to terms with Darwin. They had two approaches, each of which was in its own way sensible. One was to say that the Genesis story was a metaphor and every day represented millions of years. The other came from [Alfred Russel] Wallace—that the six days were real, but they were the days in which God put into humankind, uniquely, a sort of post-biological soul, which didn't need genes and didn't leave fossils. Most religious people are happy to accept that, and the Pope himself has recently come up with a similar claim. Not until the '60s did hard-line creationism come back to life, and mainly in the States. Why that should be isn't clear to me at all. It has a political agenda, in that most creationists are on the right and wish to believe that there is a conspiracy by the left against them: If people of liberal persuasion believe in evolution, then evolution must be wrong. But of course science, any science, isn't like that; it doesn't matter who believes in it; what matters is if it's true or not. And I have to say, evolution is true, never mind what millions might think. But why there's been a sudden outburst of antirational mania I don't understand. Maybe only an American can understand it.

EDGE: How has the evolutionary idea itself evolved?

JONES: It has, in fact, evolved in some quite unnecessary ways, because if you look back on many of the evolutionary controversies of the last 30 years, they have ebbed away as knowledge has grown. Take punctuated equilibrium, which was a useful controversy, as it made biologists feel less smug about their understanding of evolution. Or the lengthy argument about co-adaptation—the idea that genes didn't work as individual particles but as harmoniously interacting universes, and that this slowed evolution down because it was hard to get from one point to another. Or Sewall Wright's great idea that most evolution happened by accident, when you went through small bottlenecks, simply because natural selection could never get from one form to a new one without going through maladaptive forms on the way.

Most of what seemed inexplicable fits, we can now see, into orthodox Darwinian theory. People have been inspecting Darwin's feet for signs of clay since the day he died. And although some traces have been found, what's amazing is how well his edifice has lasted. Most of these evolving evolutionary ideas have gone extinct, while the original one flourishes. The only important evolutionary piece missing in 1859 was the mechanism of inheritance, but once that appeared, the edifice became so sound that much of what we've been arguing about has probably been fairly irrelevant.

EDGE: What is there about the public's understanding of science that can lead to Kansas and creationism?

JONES: The better the scientist, the narrower the mind, is a good general rule. And that's what science is; it's a collation of narrow minds all put together. Occasionally we get a more open thinker—and I would put Gould into that category—who can see a pattern missed by the broad church of cramped imaginations. The great problem with the public understanding of science, and this shows in Kansas as much as anywhere else, is not to see that. People have no insight into what you might call the grammar of science, the way it works. Many people feel that because science is filled with disagreement it must be wrong. But it's not like religion, which is filled with agreement, at least within one faith. Again, unlike religion, we tend not to talk much about the stuff we agree on and concentrate on the difficulties. That, though, is a sign of strength and not weakness. Any science in which everyone agrees about everything is dead. Compare that to faith in the Bible story of creation!

EDGE: Darwinism remains a big subject in the United Kingdom. Books on Darwin are often at the top of bestseller lists, while the same books published in the United States may receive scant review attention and come and go with little notice. Is this because Darwin is a hometown boy?

JONES: There's elements of that. Darwin was on the 20-pound note. You saw his face every day. He *is* a hometown boy; he's an iconic figure in Britain. And he's really fed into everybody, not only as a wonderful scientist but as somebody having an exciting and interesting life. Any school kid can tell you what the *Beagle* voyage was. And he comes across as such an attractive character, which helps. That's why so much of the best science writing is about evolution. Where are the Goulds or Pinkers when it comes to the chemistry of chlorine? I'm sure it's just as interesting, but as far as I know there is no King of Chlorine around whom you could weave the tale.

EDGE: What do you think about the emerging field of evolutionary psychology?

JONES: I find myself a bit depressed by the whole thing—a lot of it is the bland saying the banal. Of course, certain aspects of human behavior descend with modification from the past. About half of all genes are switched on in the brain, and it's foolish to say that those genes are different from all others, they can't evolve. It's clear that we descend from social primates, and it's no accident that the worse punishment second to the death penalty is solitary confinement. If we descended from orangutans, which are rather solitary, the worst punishment would be to force someone to give a dinner party. So there is an evolutionary psychology in that sense, obviously.

But the problem is obviousness disguised as insight. There's plenty of good work on, say, the rate at which stepmothers kill their children; that's respectable social science. But I have to say I am not surprised to find that mothers love their children more than stepmothers do. But evolutionary psychologists leap on the tables and shout we've made this fantastic discovery, equivalent to the double helix—mothers love their children! I say, what? And men are more violent than women. Well, I kind of knew that already. It's true, but it's not very profound.

And then there's this huge penumbra of pseudoscience around the subject. It's what I think of as neo-creationism. In Kansas, nothing can be explained by evolution. It's wrong—that's it. To a lot of evolutionary psychologists, though, everything in human society—war, peace, rape, marriage, the lot—can be explained by the pressure to pass on genes. But if everything can be explained, then nothing can be explained. You don't need any experiments, it's in the great Darwinian Bible. I've seen evolutionary explanations of acne, of gossiping, of ballroom dancing, the lot. It's a parlor game called Name and Explain. Just as for creationists, all this needs nothing more than belief. The infantile Darwinists are in a situation where they can't lose. If you find everything in the Bible or the *Origin,* there's no point in doing science.

Evolution is to social scientists as statues are to birds. It's a convenient platform on which to drop ill-digested ideas. An odd thing about evolutionary psychology—which is what most of the public (and, as I know to my cost, quite a few book reviewers) see as the center of the science—is that it is almost absent from the practice of evolution itself. It may be talked about in psychological conferences, but it is never mentioned in evolutionary meetings. I go to dozens of them. People argue about the fossil records, about DNA, about animal behavior, about kin selection, about the nature of species—everything is open; but evolutionary psychology is, in the eyes of evolutionists, more or less a dead duck. I've never seen any of its supporters at a scientific meeting about evolution, either. There's a kind of parallel Darwinian universe in the arts faculty out there—and I don't think the arts faculty has much useful to say about science.

EDGE: What are you trying to accomplish with *Darwin's Ghost*?

JONES: I'm doing it in part for Dr. Johnson's reason, which is that no man in his right sense has ever written except for money. But I also did it because there was a gap that needed to be filled. There's a lot of good writing about evolutionary biology, but there's no good single book about evolution. Gould writes passionately and well about fossils; Dawkins about natural selection; Pinker about behavior; Diamond about our own biological past—but every one of those topics is only a small part of the story of evolution, and only one chapter (and for humans, not even that) in *Origin of Species*.

Many years ago some colleagues and myself thought we might write an evolution text. And I had the bright idea as to how to write it—it was to take *Origin* and say, "OK, this is what the story is; why don't we just use the same logic and put modern facts in." As soon as we started, it became obvious that this was not going to be a book but a library; it would be enormous, everything from Aristotle to Zoos. So it sat in the back of my mind for 20 years, and then I started doing it on a much smaller scale. The big problem was to know what to leave out. However, it amazed me how well the structure of the original *Origin* holds up. It has a narrative flow and a structure, and all the discoveries of today bolt onto it extraordinarily well.

EDGE: Where do you see the biological sciences going in the near term?

JONES: The immediate future is one of introspection. There's already the flapping of wings as the molecular chickens come home to roost. Five years ago the optimists were saying that we will soon cure genetic disease. They've been remarkably silent the last 12 months, and they're going to be a lot more silent two years from now. The great unachievable, the human genome sequence, isn't actually answering many questions. Instead, it's asking them. And the hope of an immediate payoff is also too optimistic. The heart was dissected in 1540, the circulation of the blood was in 1670 or so, William Harvey; but the first heart transplant was in 1966. I'm not going to say it will take 400 years between the human genome sequence and the medical application of genetics, but it will take an awful lot longer than anybody had hoped. I do feel that what biology needs to do now is to sit down and think.

EDGE: What's next in terms of your own scientific research?

JONES: To get back to being one of the great narrow minds of the century. One of the things on which Gould and I are in perfect harmony is that we are each among the six top experts in the world on the genetics of snails; and the other four agree. I have plans to go back to studying the population genetics of land snails in the Pyrenees, which is a lot more fun than writing books.

6
A United Biology

E. O. Wilson

[May 26, 2003]

E. O. Wilson is Honorary Curator in Entomology, Museum of Comparative Zoology, and Pellegrino University Research Professor, Emeritus, Harvard University.

Introduction by Steven Pinker

Some sixty years ago the molecular structure of DNA was discovered and a new academic specialty came into existence. Though it was called "molecular biology," it was very different from the field that traditionally was called biology and that most people think of when they hear the word. Today the split is so pervasive that many universities have separate departments for molecular biology and traditional kind, which the molecular types denigrate as "birdsy-woodsy" biology.

Today no one personifies traditional biology more than E. O. Wilson. For nearly 60 years he has fought to unify it, revitalize it, and keep it in the public eye. The public may think of "ecology" as a romantic movement to save charismatic mammals, but it was Wilson's pioneering studies of island biogeography that helped to make it a rigorous science. Most people today consider it obvious that humans have a nature as well as a history, and that the study of our species cannot be conducted in ignorance of evolutionary biology. But it was far from obvious when Wilson first advocated that idea in 1975, at considerable personal cost. Nor should it be shocking to think that all human knowledge is connected in a single web of explanation, but it took Wilson to give this idea a name—consilience—and to become its public advocate. Few people realize that the central activities of biology—classifying species and preserving specimens—have been endangered by the molecular juggernaut; Wilson is the most visible activist dedicated to saving them. Wilson has also called attention to the deep human need to be surrounded by other living things and has made it a key argument for preserving the diversity of life in the face of today's massive human-caused extinctions. And on top of all this, Wilson's most specialized research activity—the study of ants—has made the subject so familiar to the public that two full-length animated movies have relied on ant facts for their humor.

Wilson has a restless intellect and never fails to come up with interesting new ideas. This 2003 Edge interview provides more revelations on the nature of living things from the man who has personified the science that studies them.

The sociobiology wars that began in the '70s are over. The biological approach has prevailed. Yet I totally misjudged the ignorance of the reaction I would get from social scientists and political ideologues on the left on the publication of my book *Sociobiology* in 1975. It started a real controversy and revealed the widespread—in fact, in places almost universal—belief in the blank-slate mind: that is, a mind unaffected by genetic factors or biological processes that might predispose social behavior, especially, to develop in one direction or another. That view—that the mind was fully developed by learning, by experience, and by the contingencies of history—was virtual dogma in the social sciences.

It was also dogma among the far left, who held the position taken by the Marxists and by the late Soviet Union. In the 1920s the Soviet Union dropped eugenics, a kind of prerunner of sociobiology, and switched to blank-slate dogma. That was a formidable political and academic establishment on American campuses and among American intelligentsia in the 1950s, but it seems amazing to me now, looking back on the past quarter of a century, that what I wrote could be regarded as heresy. When you read *Sociobiology* now, it looks like a fairly mild foreshadowing of what was to come.

Whatever elements of the blank slate there were in evolutionary biology and psychology vanished, or began to shrink substantially, and that's been the tendency ever since. If there are blank slaters in the social sciences and humanities today—and I suppose there still are—it's hard to see how they could hold a discussion that would include anything that we actually know about how the brain works and how child development proceeds. We've had a fair amount of success in many areas of interpreting human nature in evolutionary terms. They would just have to reject the science outright, in the manner of religious dogmatists.

My interest is ants began when I was about nine years old. I had a bug period, and I just loved the idea of going on expeditions. I grew up in Alabama and north Florida, but at the time—1939-1940—my father was a government employee in Washington, DC. We lived within walking distance of the National Zoo and Rock Creek Park, and I read *National Geographic* and thought there could be no better life in this world than going on expeditions and seeing all the wonderful things I saw and heard about in that magazine.

At the age of nine I was running my own little expeditions, with jars to preserve insects, in Rock Creek Park, and I was hooked. Pretty soon I was concentrating on ants and butterflies.

Then we went back to southern Alabama, the Gulf Coast, with its magnificent fauna and flora. This was like letting me into a candy store and I never looked back.

I went to the University of Alabama and they pretty much let me do what I wanted to do. I got into the Department of Biology and had some very good, attentive professors. It was the late '40s and they paid close attention to me. I was a gangly seventeen-year-old when I first went, and I graduated at nineteen. They were used to dealing almost entirely with preparing students to go on to medical school. Here they had an authentic embryonic biologist, so I got all sorts of special attention, including my own lab space when I was a freshman—it was great.

I'm not sure you could reproduce that experience today. Science has changed a lot. For parents thinking of encouraging their children to become scientists—and especially biologists and naturalists, if the student has that inclination to start with—I would recommend liberal arts colleges, not major research institutes. Go to a major research university after you've had four years of a liberal arts college that believes in generalized training in biology, including natural history, with heavy emphasis on ecology. In the last several years, I've visited a number of really outstanding ones and the difference between them and major research universities, including my own Harvard, is striking, in terms of what it can mean to an individual student.

Most science education takes a boot camp approach or is set up to train acolytes. That's because most scientists are journeymen—they're not masters. That is to say, they're well versed and if it's a major research university they probably have some accomplishments in a narrow segment of scientific research, but basically they think like journeymen and are there to train journeymen. They don't think particularly laterally about what their field means. There are, of course, in every university and college striking exceptions, but most scientists are recognized for and advanced by the discoveries they make. The gold and silver of science is original discovery. They know they have to be involved in making an original discovery, and to do that you move along a very narrow front.

The time will come when we'll have to move education to broaden its base for everyone. That includes far more science than is now taught, on average. The best way to treat science—I've had 41 years of experience teaching beginning students at Harvard, both biology majors and non-science students, so I can speak to this—is to take it from the top down. Put the big questions to them and show them how science can or cannot answer those questions.

Ask the questions right from the beginning of the freshman class: What is the meaning of sex? Why do we have to die? Why do people grow old? What's the whole point of all this? You've got their attention. You talk about the scientific exploration of these issues and in order to understand them you have to understand something about the whole process of evolution and how the body works. You say that we're going to deal with two great principles that are the substance of biology and which you must know: One, that everything that's in the body, including the brain and the action of the mind, is obedient to the laws of physics and chemistry as we understand it. And two, that the body, the species, and life as a whole evolved by natural selection. You take it from there and explain as best we can what we know about science, recognizing that there are still unanswered questions. If you sensibly ask what the meaning of life is, you don't have to worry about science haters or mathophobes. You've got 'em.

Lately I've been circling back to the large issue of consilience, the notion that there is a unity of the sciences through a network of cause-and-effect explanations in physics, biology, and even the lower reaches of the social sciences. To that end, in addition to doing systematic, basic biodiversity research, I'm conducting a reexamination of the basic theory and contents of sociobiology, beginning with insects and eventually coming back to humans.

In sociobiology, the social insects—ants, bees, wasps, termites—are so especially congenial to analysis, experiments, and theory that we can find paradigms of this kind of explanation that range from the genome, through the organism, through the colony, and through the ecosystems in which colonies live. By enriching the databases of each of those biological levels of organization, and developing middle-level theory in concert with that data accumulation as we go along, we can get a much clearer and quicker picture through the social insect of how social behavior evolved in the higher vertebrates.

We can define how this works by considering ant, termite, wasp, and bee colonies as superorganisms. A superorganism is an aggregation of highly organized individuals into colonies. In the case of the social insects we have a set of criteria that we use called eusociality, which has three criteria. First, there are two major castes—a queen, or sometimes a king, which constitutes a reproductive caste, and workers that don't reproduce as much if at all. Second, you have generations of grown, mature adults living with other grown mature individuals in the same community. And finally, you have mature adults that take care of the young. Those three elements are the primary criteria of what makes an advanced insect colony.

Insect eusocial colonies are superb systems that most people find intrinsically interesting. However, they are also superb study objects for the evolution of social existence. A lot of things have been happening for the last 20 years in experimental research on social organization, division of labor, communication, and genetic evolution, so the time has come for a new synthesis. I did one in 1971, pulling together most of what we knew about insect societies at that time, and rebuilt explanatory systems on the foundation of population biology. This was the beginning of sociobiology. I defined sociobiology then as the systematic study of the biological behavior of the social behavior in all kinds of organisms, and suggested that the way to make a real science of it was to base it on the study of the biology of population, recognizing that a society is a little population. This worked out very well. I also did a new synthesis called *The Ants* with Bert Hölldobler in 1990, and we are now reexamining everything based on some remarkable things we have learned in the last 10 years.

We're beginning to get some revolutionary new ideas about how social behavior originated, and also how to construct a superorganism. If we can define a set of assembly rules for superorganisms, then we have a model system for how to construct an organism. How do you put an ant colony together? You start with a queen ant, which digs a hole in the ground, starts laying eggs, and goes through a series of operations that raise the first brood. The first brood then goes through a series of operations to breed more workers, and before long you've got soldier ants, worker ants, and foragers, and you've got a teeming colony. That's because they follow a series of genetically prescribed rules of interaction, behavior, and physical development. If we can fully understand how a superorganism is put together, we'll come much closer to general principles of how an organism is put together. There are two different levels—the cells put together to make an organism, organisms put together to make a superorganism. Right now I'm examining what we know to see if there are rules of how superorganisms are put together.

Superorganisms are superior as an experimental object, because you can do experiments in the laboratory much faster with a bunch of ants. You can take a group of ants, divide them into ten parts and experiment with them as ten parts. Suppose I were working with the operation of your hand. I could do an experiment in which I painlessly and bloodlessly cut off eight of your fingers and see how you work with two, and then put all the others back on. That's what you can do with an ant colony much more easily. You can separate workers from the colony to experiment, put them back together, and so on.

We're moving rapidly in this area. A little less than 50 years ago, shortly after the discovery of the structure of DNA—one of the epochal events of science—Jim Watson, one of the first of the newly defined full-blooded molecular biologists, came to Harvard.

Jim and I were assistant professors together at the time, and participated in a clash of civilizations. Jim, of course, was leading the molecular revolution, and for the time being I was a distinctly overmatched younger leader in organismic biology. For a couple of decades thereafter, molecular biology proceeded in its own way and began to send tendrils of investigation up into cell biology, now organismic biology, past the level of the genome in terms of truly major new discoveries, and into the great Pacific Ocean, so to speak, of proteomics, the science of how proteins are assembled.

Meanwhile, evolutionary biology and organismic biology continued to grow in strength and sophistication and extended their reach on down beyond the organism. By the 1980s, we were learning about the genome, the molecular biology itself, and the consequence has been, by the late '80s and '90s and now increasingly, that these two once distant levels are pretty well connected, and people are moving back and forth across them rather easily. Increasingly the study of biological diversity—the variety of life and how it originated—is coming to occupy the attention of even the molecular biologists.

In that spirit of solidifying the newly found unity of the different levels of biology from ecosystems, organisms, and society down to the molecular level of genomics, and with this newfound confidence that we can do this, biology is becoming a unified, mature science, and we now find that the old conflict's gone. Jim and I will be having a public dialogue soon on the relation between DNA and the great discoveries in the molecular period on the one side, and the exploration of the world's biodiversity on the other. We can put those things together in discussion now, and this will be an interesting conversation, I hope. At the very least it's symbolic of how much has happened in this half-century in biology since we started here in the department at Harvard as adversaries.

At the same time, however, biology is far from being a fully mature science. A mature science would be one in which we thoroughly understand the following big, open topics:

One is the nature of consciousness and of mind. These are biological subjects, and they're phenomena not just limited to human beings, since we can see their early origins in other vertebrates, particularly the other primates.

Another principal domain in biology that is still largely unexplored is the assembly and maintenance of ecosystems. How do ecosystems—assemblages of plants and animals—live more or less stably for an indefinite period of time? How do they come together in the first place? How are certain species chosen to enter that community? How do they manage to survive? And how does the ecosystem fit together in a way that provides stability?

We're nibbling at the edges of these issues, but community ecology is far from getting out of its infancy. This is still an open question of primary importance not only for the biology of the whole but, of course, for the sustainable use of our resources and for the saving of the rest of life through scientifically based conservation.

Conceptually, the development of a united biology would also certainly include what we're calling proteomics. This relates to the question of how, after elementary transcription and formation of the proteins, genes are turned on and off. They appear and then do certain things, in part, due to context, location, and preexisting proteins. It takes 100,000 or 200,000 kinds of proteins to form a cell.

How exactly do they come together? Most molecular biologists are now focusing on that area. Right now we're at the level of hundreds of species whose genomes have been decoded pretty thoroughly. Once we understand more about the diversity in the genetic code of thousands of species, what strategies will we be able to see the genes following as they create proteins and as the proteins assemble cells? What pathways of evolution have been followed in the course of making what adaptations in the environment?

This brings us to the entire question of biological diversity, which is one of my major concerns now. We probably know, and in the most elementary manner, no more than 10 percent of the species of plants, animals, and microorganisms sufficiently well enough to give the species a scientific name. Ninety percent remain unknown, particularly if you throw in the bacteria and other similar, simple organisms called archaeons. We've got to get about the business of exploring the planet's biodiversity.

The project I've put my shoulders behind was called the All-Species Project. We had a summit conference here at Harvard a little less than two years ago to bring together people who want to see this happen and believe it could be made to happen in 25 years, much in the manner of the human genome project, if we were to want it. We agreed on the main issue: that new technologies make it possible for this to be done. We're ready to build up the systematics of biodiversity exploration around the world and could pull this off and make a huge difference in biology and environmental management.

We established enthusiastic partnerships with various agencies, both governmental and non-governmental, but the sinking economy brought us up against the wall of insufficient funding. There are still major enterprises around the world that are doing this on a continental level, or are starting up on global scale, so within a relatively short period of time we'll see these efforts begin to coalesce. I just got off the phone giving a positive evaluation of the fund-raising drive of one of these organizations, which is going for a $9 million endowment. They're getting ambitious, have a compelling argument, and I believe it will happen. It's just a matter of when. The All-Species Project is simply a term used to describe this worldwide movement to complete the exploration of the planet.

If we do not know 90 percent of the kinds of organisms that exist on Earth, what would knowing almost all mean for us? There are overpowering arguments for undertaking this project. It would mean that for the first time we would know all of the bacteria around the world. We would understand potential disease organisms, as well as the fundamental bacterial elements of ecosystems, the primitive but elementary organisms that form a large part of the base of the ecosystem. Right now we don't even know what the majority of organisms are doing. After cataloguing all of the world's species we would have a huge reservoir of knowledge from which to draw genes for transgenic changing of crops and the development of new pharmaceuticals.

This would also greatly enlarge biology. Bear in mind that biology is primarily a descriptive science. It deals with the particularity of species and their adaptation to the environment. Although biology is based on the principles of physics and chemistry—at least it is consistent with them—its actual substance is an account of the individual biologies of thousands and eventually millions of species, each of which has a unique history, in many cases millions of years old, of exquisitely adapting and interacting with one another at certain parts of the environment. We don't know what most of that is. Supporting All Species' effort for the planet has not only a logic behind it—to bring biology more quickly to maturity—but also promises huge practical applications.

I talked about some of these issues in my book *The Future of Life*, which came out early in 2002 and had some success, but there has to be a change in the culture. The way this can be accomplished is to open the eyes of the scientific community, the government, and non-governmental supporting organizations to the tremendous cost-effectiveness and potential benefits of pulling off a full inventory of species-level diversity.

For the past 20 years, during my service on the boards of directors or advisory boards of most of the major global conservation organizations and in my research in this field, there has been a question of how to balance the importance of saving biodiversity with that of saving jobs and helping the poor. There isn't any question anymore that saving the rest of life is compatible with saving and improving the lot of humanity. In fact, to push for one means to push for the other, and to let the one go means that you let a lot of the other go.

The major global conservation organizations have long since included in their programs an emphasis on economic development, on-site pilot programs, and fund-raising to improve economies in areas of high conservation value on a sustainable basis. And it turns out that it works. I could spend hours talking about the examples and the economics of it and so on, but the bottom line is that the two great goals of the 21st century are, first, raising people around the world to a decent standard of living, particularly the 80 percent of the people living in developing countries, and second, bringing as much of the rest of life through with us. If we can do this, we will obtain the kind of better world that people everywhere believe should be our major human purpose.

And it's practicable—it is not at all expensive, in terms of the world domestic product. For example, Conservation International convened economists and biologists two years ago in order to estimate how much it would cost to save the rest of biodiversity. It turns out that in order to save the world's 25 hottest hot spots—those places where you have the greatest endangerment to whole ecosystems with large numbers of species—and then add the cost of saving the core

wilderness areas of the great tropical forests of the Congo, the Amazon, and New Guinea, it would cost one payment of about $28 billion.

This is equal to approximately one part in a thousand of the world's domestic product. That's one-tenth of one percent of the annual economic output of the world! One payment could cover 70 percent of the species of plants and animals we know about on Earth, so this is something that's obtainable. And part and parcel of that would be to improve the economies of the areas in which the main biodiversity is located.

7
Is Life Analog Or Digital?

Freeman Dyson

[March 13, 2001]

Freeman Dyson is a theoretical physicist at the Institute for Advanced Study

One of my favorite books is *Great Mambo Chicken and the Transhuman Condition* by Ed Regis. The book is a collection of stories about weird ideas and weird people. The transhuman condition is an idea suggested by Hans Moravec. It is the way you live when your memories and mental processes are downloaded from your brain into a computer. The wiring system of the computer is a substitute for the axons and synapses of the brain. You can then use the computer as a back-up, to keep your personality going in case your brain gets smashed in a car accident, or in case your brain develops Alzheimer's. After your old brain is gone, you might decide to upload yourself into a new brain, or you might decide to cut your losses and live happily as a transhuman in the computer. The transhumans won't have to worry about keeping warm. They can adjust their temperature to fit their surroundings. If the computer is made of silicon, the transhuman condition is silicon-based life. Silicon-based life is a possible form for life in a cold universe to adopt, whether or not it happens to begin with water-based creatures like us made of flesh and blood.

Another possible form of life is the Black Cloud described by Fred Hoyle in his famous science fiction novel. The Black Cloud lives in the vacuum of space and is composed of dust-grains instead of cells. It derives its energy from gravitation or starlight and acquires chemical nutrients from the naturally occurring interstellar dust. It is held together by electric and magnetic interactions between neighboring grains. Instead of having a nervous system or a wiring system, it has a network of long-range electromagnetic signals that transmit information and coordinate its activities. Like silicon-based life and unlike water-based life, the Black Cloud can adapt to arbitrarily low temperatures. Its demand for energy will diminish as the temperature goes down.

Silicon-based life and dust-based life are fiction and not fact. I use them as examples to illustrate an abstract argument. The examples are taken from science fiction but the abstract argument is rigorous science. The abstract concepts are valid, whether or not the examples are real. The concepts are digital-life and analog-life. The concepts are based on a broad definition of life. For the purposes of this discussion, life is defined as a material system that can acquire, store, process, and use information to organize its activities. In this broad view, the essence of life is information, but information is not synonymous with life. To be alive, a system must not only hold information but process and use it. It is the active use of information, and not the passive storage, that constitutes life.

The two ways of processing information are analog and digital. An LP record gives us music in analog form, a CD gives us music in digital form. A slide-rule does multiplication and division in analog form, an electronic calculator or computer does them in digital form. We define analog-life as life that processes information in analog form, digital-life as life that processes information in digital form. To visualize digital-life, think of a transhuman inhabiting a computer. To visualize analog-life, think of a Black Cloud. The next question that arises is, are we humans analog or digital? We don't yet know the answer to this question. The information in a human is mostly to be found in two places, in our genes and in our brains. The information in our genes is certainly digital, coded in the four-level alphabet of DNA. The information in our brains is still a great mystery. Nobody yet knows how the human memory works. It seems likely that memories are recorded in variations of the strengths of synapses connecting the billions of neurons in the brain with one another, but we do not know how the strengths of synapses are varied. It could well turn out that the processing of information in our brains is partly digital and partly analog. If we are partly analog, the downloading of a human consciousness into a digital computer may involve a certain loss of our finer feelings and qualities. That would not be surprising. I certainly have no desire to try the experiment myself.

There is a third possibility: that the processing of information in our brains is done with quantum processes, so that the brain is a quantum computer. We know that quantum computers are possible in principle, and that they are in principle more powerful than digital computers. But we don't know how to build a quantum computer, and we have no evidence that anything resembling a quantum computer exists in our brains. Since we know so little about quantum computing, I do not consider it in this discussion.

I started thinking about the abstract definition of life twenty years ago, when I published a paper in *Reviews of Modern Physics* about the possibility that life could survive forever in a cold expanding universe. I proved to my own satisfaction that survival is possible for a community of living creatures using only a finite store of matter and energy. Then, two years ago, Lawrence Krauss and Glenn Starkman, friends of mine at Case Western

Reserve University, sent me a paper with the title "Life, the Universe, and Nothing." They say flatly that survival of life forever is impossible. They say that everything I claimed to prove in my *Reviews of Modern Physics* paper is wrong. I was happy when I read the Krauss-Starkman paper. It is much more fun to be contradicted than to be ignored.

In the two years since I read their paper, Krauss and Starkman and I have been engaged in vigorous arguments, writing back and forth by e-mail, trying to pokes holes in each other's calculations. The battle is not over, but we have stayed friends. We have not found any holes that cannot be repaired. It begins to look as if their arguments are right, and my arguments are right too. We can both be right because we're making different assumptions about the nature of life. It turns out that they are right, and life cannot survive forever if life is digital, but I am right, and life may survive forever if life is analog. This conclusion was unexpected. In the development of our human technology during the last fifty years, analog devices such as LP records and slide-rules appear primitive and feeble, while digital devices are overwhelmingly more convenient and powerful. In the modern information-based economy, digital wins every time. So it was unexpected to find that under very general conditions, analog life has a better chance of surviving than digital life. Perhaps this implies that when the time comes for us to adapt ourselves to a cold universe and abandon our extravagant flesh-and-blood habits, we should upload ourselves to black clouds in space rather than download ourselves to silicon chips in a computer center. If I had to choose, I would go for the black cloud every time.

The superiority of analog-life is not so surprising if you are familiar with the mathematical theory of computable numbers and computable functions. Marian Pour-El and Ian Richards, two mathematicians at the University of Minnesota, proved a theorem twenty years ago that says, in a mathematically precise way, that analog computers are more powerful than digital computers. They give examples of numbers that are proved to be noncomputable with digital computers but are computable with a simple kind of analog computer. The essential difference between analog and digital computers is that an analog computer deals directly with continuous variables while a digital computer deals only with discrete variables. Our modern digital computers deal only with zeroes and ones. Their analog computer is a classical field propagating though space and time and obeying a linear wave equation. The classical electromagnetic field obeying the Maxwell equations would do the job. Pour-El and Richards show that the field can be focused on a point in such a way that the strength of the field at that point is not computable by any digital computer but it can be measured by a simple analog device. The imaginary situation they consider has nothing to do with biological information. The Pour-El–Richards theorem does not prove that analog-life will survive better in a cold universe. It only makes this conclusion less surprising.

The argument of Krauss and Starkman is based on quantum mechanics. If any material system, living or dead, is finite, it will have only a finite set of accessible quantum states. A finite subset of these states will be ground states with precisely equal energy, and all other states will have energies separated from the ground states by a finite energy gap. If the system could live forever, the temperature would ultimately become much lower than the energy gap, and the states above the gap would become inaccessible. From that time on, the system could no longer emit or absorb energy. It could store a certain amount of information in its permanently frozen ground states, but it could not process the information. It would be, according to our definition, dead. Krauss and Starkman thought they had dealt a fatal blow to my survival strategy with their argument. But I am still on my feet, and here is my rebuttal. Their argument is valid for any system that stores information in devices confined within a volume of fixed size as time goes on. It is valid for any system that processes information digitally, using discrete states as carriers of information. In a digital system, the energy gap between discrete states remains fixed as the temperature goes to zero, and the system ceases to operate when the temperature is much lower than the energy gap. But this argument does not apply to a system based on analog rather than digital devices. For example, consider a living system like Hoyle's Black Cloud, composed of dust grains interacting by means of electric and magnetic forces. After the universe has cooled down, each dust grain will be in its ground state, so that the internal temperature of each grain is zero. But the effective temperature of the system is the kinetic temperature of random motions of the grains. Since electric and gravitational energies vary inversely with distance, the cloud must expand as its temperature cools. A simple calculation shows that in spite of the falling temperature, the number of quantum states accessible to each grain increases with the three-halves power of the size of the cloud. The number of quantum states grows larger and larger as the cloud expands. In an analog system of this kind, there is no ground state and no energy gap.

An analog form of life, such as Hoyle's Black Cloud, adapts better to low temperatures, because a cloud with a fixed number of grains can expand its memory without limit by increasing its linear scale. The quantized-energy argument does not apply to an analog system, because the number of quantum states is unbounded. At late times, quantum mechanics becomes irrelevant, and the behavior of the system becomes essentially classical. The number of quantum states becomes so large that classical mechanics becomes exact. When analog systems work classically, the quantized-energy argument fails. That is why survival is possible in the domain of classical mechanics although it is impossible in the domain of quantum mechanics. Fortunately, classical mechanics becomes dominant as the universe expands and cools. But Krauss and Starkman have not yet conceded. I am still expecting them to come back with new arguments, which I will then do my best to refute.

It seems to me now that the question of whether life is analog or digital is more interesting, and perhaps more important, than the question of ultimate survival out of which it arose.

8
Life: What A Concept!

An Edge Special Event at Eastover Farm, August 27, 2007
Speakers:

Freeman Dyson, theoretical physicist, Institute for Advanced Study

J. Craig Venter, leading genomics scientist;cofounder & chairman, Synthetic Genomics

George Church, professor of genetics, Harvard Medical School

Dimitar Sasselov, professor of astronomy, Harvard; director, Harvard Origins of Life Institute

Seth Lloyd, quantum mechanical engineer, MIT

John Brockman, publisher of Edge and owner of Eastover Farm

FREEMAN DYSON: First of all, I want to talk a bit about the origin of life. To me the most interesting question in biology has always been how it all got started. That has been a hobby of mine. We're all equally ignorant, as far as I can see. That's why somebody like me can pretend to be an expert.

I was struck by the picture of early life that appeared in Carl Woese's article ["A New Biology for a New Century"] three years ago. He had this picture of the pre–Darwinian epoch, when genetic information was open-source and everything was shared between different organisms. That picture fits nicely with my speculative version of the origin of life.

The essential idea is that you separate metabolism from replication. We know modern life has both metabolism and replication, but they're carried out by separate groups of molecules. Metabolism is carried out by proteins and all kinds of small molecules, and replication is carried out by DNA and RNA. That may be a clue to the fact that they started out separate rather than together. So my version of the origin of life is that it started with metabolism only.

You had what I call the garbage-bag model. The early cells were just little bags of some kind of cell membrane, which might have been oily or it might have been a metal oxide. And inside you had a more-or-less random collection of organic molecules, with the characteristic that small molecules could diffuse in through the membrane but big molecules could not diffuse out. By converting small molecules into big molecules, you could concentrate the organic contents on the inside, thus the cells would become more concentrated and the chemistry would gradually become more efficient. So these things could evolve without any kind of replication. It's a simple statistical inheritance. When a cell became so big that it got cut in half, or shaken in half by some rainstorm or environmental disturbance, it would then produce two cells, which would be its daughters and would inherit, more or less, but only statistically—the chemical machinery inside. Evolution could work under those conditions.

LLOYD: These are naturally occurring lipid membranes?

DYSON: Yes. Which we do know exist. That's stage one of life, this garbage-bag stage, where evolution is happening but only on a statistical basis. I think it's right to call it pre–Darwinian, because Darwin himself did not use the word "evolution"; he was primarily interested in species, not in evolution as such.

Well then, what happened next? Stage two is when you have parasitic RNA—when RNA happens to occur in some of these cells. There's a linkage, perhaps, between metabolism and replication in the molecule ATP [adenosine triphosphate]. We know that ATP has a dual function: It's important for metabolism, but it also is essentially a nucleotide—you only have to add two phosphates and it becomes a nucleotide. So it gives you a link between the two systems. Perhaps one of these garbage bags happened to develop ATP by a random process. ATP is helpful to the metabolism, so these cells multiplied and became very numerous and made large quantities of ATP. Then by chance this ATP formed the adenine nucleotide, which polymerized into RNA. You had, then, parasitic RNA inside these cells—a separate form of life, which was pure replication without metabolism. RNA could replicate itself. It couldn't metabolize, but it could grow quite nicely.

Then RNA invented viruses. RNA found a way to package itself in a little piece of cell membrane and travel around independently. Stage two of life has the garbage bags still unorganized and chemically random, but with RNA zooming around in little packages we call viruses, carrying genetic information from one cell to another. That's my version of the RNA world. It corresponds to what Manfred Eigen considered the beginning of life, and which I regard as stage two. You have RNA living independently, replicating, traveling around, sharing genetic information between all kinds of cells. Then stage three, which I would say is the most mysterious, began when those two systems started to collaborate. It began with the invention of the ribosome, which to me is the central mystery. There's a tremendous lot to be done with investigating the archaeology of the ribosome. I hope some of you people will do it.

Once the ribosome was invented, then the two systems—the RNA world and the metabolic world—are coupled together and you get modern cells. That's stage three, but still with the genetic information being shared, mostly by viruses traveling from cell to cell, so it's open-source heredity. As Carl Woese described it, evolution could be very fast.

That's roughly the situation as Carl Woese described it: You have modern cells with metabolism directed by RNA or DNA but without any private intellectual property, so that the chemical inventions made by one cell could be shared with others. Evolution could go in parallel in many different cells, so it could go a lot faster. The best chemical devices could be shared between different cells and combined, so evolution would go rapidly in parallel. That was probably the fastest stage of chemical evolution, when most of the basic biochemical inventions were made.

Stage four is the stage of speciation and sex, which are the next two big inventions, and that's the beginning of the Darwinian era, when species appeared. Some cells decided it was advantageous to keep their intellectual property private—to have sex only with themselves or members of their own species, thereby defining species. That was then the state of life for the next 2 billion years, the Archaeozoic and Proterozoic eras. It was a rather stagnant phase of life and continued for 2 billion years without evolving fast.

Then you had stage five, the invention of multicellular organisms, which also involved death, another important invention.

Then after that came us—stage six. That's the end of the Darwinian era, when cultural evolution replaces biological evolution as the main driving force. "Cultural" means that the big changes in living conditions are driven by humans spreading their technology and their ways of making a living, by learning from one another rather than by breeding. So you are spreading ideas much more rapidly than you're spreading genes. And stage seven is what comes next.

The question is whether any of that makes sense. I think it does, but like all models it's going to be short-lived and soon replaced by something better.

The other thing I was going to talk about was domesticated biotech, which is a completely separate subject. That comes from looking around at what's happened to physics technology in the last 20 years, with things like cell phones and iPhones and the things I see around me at the table. Personal computers of all kinds. Digital cameras. And the GPS navigation system. All those wonders of technology, which have suddenly descended from the sky to the earth. They have become domesticated. That has been a tremendous change, something we never predicted.

I remember when John von Neumann was developing the first programmable computer at Princeton. I happened to be there, and he talked a lot about the future of computing, and he thought of computers as getting bigger and bigger and more and more expensive, so they belonged to big corporations and governments and big research labs. He never in his wildest dreams imagined computers being owned by three-year-olds and being part of the normal upbringing of children. It's said that somebody asked him at one point how many computers would the United States need? How large would the market be? And he answered, "Eighteen." It went in totally the opposite direction.

VENTER: Well, it went in both directions.

DYSON: To some extent, but even the biggest computers are not much bigger than they were in those days. It's remarkable. I remember the very first computer in Princeton; it was a huge thing, a room about as big as this tent, full of machinery. This was in 1951, '52. It was running smoothly around '53.

VENTER: But it was less powerful than your laptop.

DYSON: Oh, much less! The total memory was 4 kilobytes. And he did an amazing lot with that. Especially a mathematician who was there at the time, Nils Barricelli, did simulated evolution amazingly well with a memory of 4 kilobytes. He developed models of evolving creatures forming an ecology, and they showed punctuated equilibrium, exactly the way real species do. It was astonishing how much he could get out of that machine.

LLOYD: The problem is that computers get faster by a factor of 2 every year and a half, but computer programmers conspire to make them run slightly slower every year and a half, by junking them up with all sorts of garbage.

DYSON: Because von Neumann thought he was dealing with unreliable hardware, he made another mistake. The problem was how to write reliable software so as to deal with unreliable hardware. Now we have the opposite problem: Hardware is amazingly reliable, but software is not. It's the software that sets the limit to what you can do.

My prediction is that the same thing is going to happen to biotech in the next 50 years, perhaps 20 years—that it's going to be domesticated. And I take the example of the flower show in Philadelphia and the reptile show in San Diego, at both of which I saw demonstrations of the enormous market there is for people who are skilled breeders of plants and animals. And they're itching to get their hands on this new technology. As soon as it's available I believe it's going to catch fire, the way computers did when they became available to people like you. It's essentially writing and reading DNA, breeding new kinds of plants and trees and bushes by writing the genomes at home on your personal machine. Just a little DNA reader and a little DNA writer on your desk, and you play the game with seeds and eggs instead of with pictures on the screen. That's all.

LLOYD: One of the reasons computers became ubiquitous is the phenomenon of Moore's Law, where they became faster and more powerful by a factor of 2 every 2 years. Is there an equivalent here?

DYSON: Exactly the same thing is happening to DNA at the moment. Moore's Law is being followed as we speak, both by reading and writing machines.

LLOYD: At roughly the same rate?

DYSON: Yes.

VENTER: It's happening faster. I had this discussion with Gordon Moore, and I said that sequence-reading and writing was changing faster than Moore's Law, and he said, "But it won't matter, as you're ultimately dependent on Moore's Law."

DYSON: I agree with that. At the moment, it's going fast.

CHURCH: Unless we build bio-computers—right now the best computers are bio-computers.

BROCKMAN: It took 2 weeks for a seventeen-year-old to hack the iPhone, and here we're talking about DNA writers and readers. That same kid is going to start making people.

DYSON: That's true; the driving force is the parents, not the scientists. Fertility clinics are a tremendously large and profitable branch of medicine, and that's where the action is. There's no doubt this is going into fertility clinics as well. For good or evil, that's happening.

BROCKMAN: But isn't this a watershed event because of our ideas about life? What's possible will happen. What will the societal impact be?

DYSON: It's not true that what's possible will happen. We have strict laws about experimenting with human subjects.

BROCKMAN: You can't hack an iPhone either; certain activities along these lines are illegal.

DYSON: But it's different with medicine. You do get put in jail if you break the rules. There are clear similarities but also great differences. Certainly it is true that people are going to be monkeying around with humans; I totally agree with that. But I think that society will put limits on it, and that the limits are likely to be broken from time to time but they will be there.

SHAPIRO: I just want to bring in one distinction here, because two things are getting confused. To go to computers, I remember that perhaps 30 years ago there was something called a Heathkit and the idea was, Why buy a computer when you can build your own computer in your basement? Well, I don't see anyone constructing their own computer in their basement anymore. If you purchase computers from Dell or IBM, they will assemble them for you. But the actual construction, the difficult part, takes place in specialized institutions and then they make their products available. Everyone has a cell phone, but I doubt that most people, if they dropped it, could repair their cell phone. And the new biotechnology— DNA sequencing will be done massively, and engineering will be done massively, and new organisms will be constructed. But they'll be done in specialized facilities. Only the products will become available to the general public. No child will go into his basement and set up the necessary DNA synthesizers, or DNA sequencers, and proceed to make his own new organisms.

DYSON: You're thinking like von Neumann, and I disagree. What you'll sell to the kids is kits. You won't sell the whole apparatus for doing things, but you'll sell a kit that will do the things that are fun, just as you do with computers that are sold for children to play games. The computers only play games, they don't actually calculate numbers.

LLOYD: In fact theres a good analogy in the history of computation. Thirty years ago, MIT freshmen arrived having built a computer, and then shortly after that they stopped building computers. Twenty years ago, or 15 years ago, they arrived knowing how to program computers. But nowadays when freshmen arrive, far fewer of them have actually programmed a computer before, in the sense of writing a program in a language such as Java. But they use computers far more, and they're great users of software. They know vast amounts about how computers work and what you can do with the software. Why? Because it's a lot easier to do. Why program a computer if somebody can enable you to just use the software and program it? Of course, when you're playing Grand Theft Auto, you're effectively programming the computer at the same time. So I suspect that what Freeman says is right: People will be using this new genetic technology, but maybe there's an analog of programming in the constructing of new organisms which will enable people to do it—an analog of software so people will become the users of the software.

SHAPIRO: I see children being able to buy lizards, say, that glow in the dark with green fluorescence, but I don't see them creating them in their basement.

DYSON: I think both are going to happen.

SASSELOV: Maybe the question is, What is the time scale for the second thing happening? That is, by then the technology will be so developed that we may be different as a species and not care as much as we do today whether some kid is capable of tinkering with a human. Because we will have tinkered enough, in the regulated way, by then, so that it wouldn't matter as much.

DYSON: Yes, nobody can ever know in advance; all these things always turn out differently than you expected.

LLOYD: In fact this is a real specter, because, as you say, we're not allowed to tinker with humans but we are allowed to tinker with rats—that we very rapidly will develop rats that surpass us in all abilities, whereas we're just stuck in the dark ages.

BROCKMAN: Freeman, last night I asked Richard Dawkins if he cared to comment on your article suggesting "the end of the Darwinian Interlude." ["Our Biotech Future," *The New York Review of Books*, July 19, 2007] He sent the following comment, with the caveat that it's a hastily written response solely for the purpose of this meeting. He writes:

> " '*By Darwinian evolution he [Woese] means evolution as Darwin understood it, based on the competition for survival of noninterbreeding species.' [and] 'With rare exceptions, Darwinian evolution requires established species to become extinct so that new species can replace them.*' These two quotations from Dyson constitute a classic schoolboy howler, a catastrophic misunderstanding of Darwinian evolution. Darwinian evolution, both as Darwin understood it, and as we understand it today in rather different language, is not based on the competition for survival of species. It is based on competition for survival within species. Darwin would have said competition between individuals within every species. I would say competition between genes within gene pools. The difference between those two ways of putting it is small compared with Dyson's howler (shared by most laymen: it is the howler that I wrote *The Selfish Gene* partly to dispel, and I thought I had pretty much succeeded, but Dyson obviously hasn't read it!) that natural selection is about the differential survival or extinction of species. Of course the extinction of species is extremely important in the history of life, and there may very well

be non-random aspects of it (some species are more likely to go extinct than others) but, although this may in some superficial sense resemble Darwinian selection, it is not the selection process that has driven evolution. Moreover, arms races between species constitute an important part of the competitive climate that drives Darwinian evolution. But in, for example, the arms race between predators and prey, or parasites and hosts, the competition that drives evolution is all going on within species. Individual foxes don't compete with rabbits, they compete with other individual foxes within their own species to be the ones that catch the rabbits (I would prefer to rephrase it as competition between genes within the fox gene pool).

"The rest of Dyson's piece is interesting, as you'd expect, and there really is an interesting sense in which there is an interlude between two periods of horizontal transfer (and we mustn't forget that bacteria still practice horizontal transfer and have done throughout the time when eucaryotes have been in the 'Interlude'). But the interlude in the middle is not the Darwinian Interlude, it is the Meiosis / Sex / Gene-Pool / Species Interlude. Darwinian selection between genes still goes on during eras of horizontal transfer, just as it does during the Interlude. What happened during the 3-billion-year Interlude is that genes were confined to gene pools and limited to competing with other genes within the same species. Previously (and still in bacteria) they were free to compete with other genes more widely (there was no such thing as a species outside the 'Interlude'). If a new period of horizontal transfer is indeed now dawning through technology, genes may become free to compete with other genes more widely yet again.

"As I said, there are fascinating ideas in Freeman Dyson's piece. But it is a huge pity it is marred by such an elementary mistake at the heart of it."

DYSON: Good. Yes, I have two responses. First, what I wrote is not a howler and Dawkins is wrong. And I *have* read his book.

Species, once established, evolve very little, and the big steps in evolution mostly occur at speciation events, when new species appear with new adaptations. The reason for this is that the rate of evolution of a population is roughly proportional to the inverse square root of the population size. So big steps are most likely when populations are small, giving rise to the "punctuated equilibrium" seen in the fossil record. The competition is between the new species with a small population adapting fast to new conditions and the old species with a big population adapting slowly.

Second, it's absurd to think that group selection is less important than individual selection. Consider for example Dodo A and Dodo B, competing for mates and progeny in the dodo population on Mauritius. Dodo A competes much better and has greater fitness, as measured by individual selection. Dodo A mates more often and has many more grandchildren than Dodo B. A hundred years later, the species is extinct and the fitness of A and B are both reduced to zero. Selection operating at the species level trumps selection at the individual level. Selection at the species level wiped out both A and B, because the species neglected to maintain the ability to fly, which was essential to survival when human predators appeared on the island. This situation is not peculiar to dodos; it arises throughout the course of evolution, whenever environmental changes cause species to become extinct.

In my opinion, both these responses are valid, but the second one goes more directly to the issue that divides Dawkins and myself.

VENTER: I have trouble with some of the fundamental terms. What's your definition of "species"? That's something I have great difficulty with, lately, out of our research.

DYSON: Yes, it is a problem. It's supposed to be just a population that breeds within the population but not outside, but of course there are all sorts of exceptions.

VENTER: That ignores most of biology.

DYSON: Yes, so I don't know what the real definition is. But that's the conventional definition.

VENTER: It's a human definition.

DYSON: It is fuzzy. Like most things.

LLOYD: So for sexually reproducing species, then, it's less fuzzy than for bacteria.

DYSON: Right.

VENTER: But it really comes down to one or two recognition molecules that determine the species. If it's based on interbreeding, it's the sperm recognition sites, right?

DYSON: Yes.

VENTER: So that determines the species, then.

DYSON: Well, amongst other things.

CHURCH: Chromosome dynamics, morphology, behavior—many things. Depending on how complex the organism is.

VENTER: It's easy to tell a human from a giraffe, and we can call that a different species.

DYSON: One of the books I've learned most from, is Jonathan Weiner's *The Beak of the Finch*, which describes evolution as it's observed in the Galapagos by [evolutionary biologists] Peter and Rosemary Grant. It's remarkable that they can actually see, from year to year, species starting to hybridize when conditions are good and then separating again when conditions are bad. So, even on a year-to-year time scale, you can actually see this happening—that species are not well-defined.

LLOYD: Sorry, I'm not familiar with this work. So they hybridize when times are good, and when times are bad they separate into smaller populations. Is this so they can evolve more rapidly?

DYSON: Yes. So they can specialize. Because in bad times you have to specialize in chewing particular seeds.

VENTER: During droughts, all that was left were these really hard seeds. Finches that survive have Arnold Schwarzenegger beaks.

DYSON: Not only those—you can also have a separate population that specializes in the small seeds, which have small beaks. It happens, because of the geography, that you have violent swings in climate. During El Niño, conditions are wet, and between El Niños conditions are dry. So selection is brutal—almost every year about half of them get selected out.

VENTER: One of the highlights of my round-the-world expedition was meeting up with the Grants in the Galapagos, in their little tent on the site of Daphne Major. They spent 3 months on this island in this little tent; there's no fresh water, there's nothing there. And they live off of bottled water and cans of tuna fish. I took them a bottle of chilled champagne. It became a happier ecosystem. Remarkable what they've done.

DYSON: The enormous advantage they had was that the birds are completely tame. You can just walk up to a bird and put a ring around its leg and it doesn't fly away. That's what made it all possible. They know every bird personally.

VENTER: Better than tame. If you walk on their path, the boobies and stuff will peck at your leg. It's their island. The humans become non-tame after a while. But, so that's an important part of the definition. Are the finches with the larger beaks a different species, in your view?

DYSON: Yes, according to Darwin they are. In fact, they do interbreed quite extensively.

VENTER: So a 2-base-pair change in a genome could be sufficient to create a new species—out of 1.5 billion.

DYSON: Yes.

VENTER: I'm not sure everybody will buy that definition. That makes you a very different species from George.

DYSON: The real problem is the lawyers. You have the Endangered Species Act; that means you have to make a legal definition of the species.

CHURCH: That's true. We're all endangered.

LLOYD: I gather human beings are a genetically non-diverse species. We take two squirrels on this tree right here—they're much farther apart genetically than we are from any other human being on the face of the Earth. So we're inclined to see things in our own light.

VENTER: What's your evidence for that?

CHURCH: It's true for chimpanzees; I don't know about squirrels.

LLOYD: But *Homo sapiens* is a quite recent species—and also the mitochondrial DNA evidence suggests that we're descended from common ancestors in the not very distant past, within the last 100,000 years or so. So there seems to have been a genetic bottleneck in the human species, compared with hominids as a whole, within the last 100,000 years. Which makes us much less diverse than, for instance, squirrels.

SHAPIRO: The thrust of what Freeman was saying—if we accept most of what he said, which I certainly do—is that concepts like species and interbreeding are about to become, in a sense, extinct. Because entering the new era, laboratories will exist which will re-create species or combine qualities of one species with qualities of another, and it will be up to the designer, the extent to which they interbreed, or interbreed with existing organisms, and so on. So that conceivably, if civilization continues, we will then be in charge of what species may come into being and what species do not.

LLOYD: I have a query: Is that actually important? Freeman, you said we reached the end of Darwinian evolution, where human beings are the dominant species on Earth, and species that can't co-evolve with humans are probably doomed. But this means that in this end of Darwinian evolution, genes are no longer so important, and instead ideas, which can be generated more rapidly—and, dare I even say, things like computations and software—are more important. Are you envisaging an era when genetic information returns to the predominant position that it had for billions of years on Earth?

DYSON: No, I don't look very far. I'm quite conservative as far as human society is concerned. We would be wise to keep ourselves as much as possible the way we are, and I hope we'll be successful in it. I don't see any great likelihood if you monkey around with humans that you'll produce anything much better.

BROCKMAN: This sounds like an engineer's approach, rather than a thinker's approach. As a scientist, aren't you talking about a huge watershed concerning our ideas of what it means to be human or even what it means to be alive? Can you imagine what ideological factions or religious groups would do with some of the statements that have been made this afternoon?

LLOYD: Ironically, many religions are sets of ideas, and one of the things that many religions tend to do is to try to sequester themselves genetically. Keep the gene pool within this religion—prevent intermarriage with people of other faiths. You could say that religion is almost an attempt by ideas to get back to the good old days of rapid evolution via genetic engineering in small populations.

DYSON: I'm not familiar with this feeling that culture is collapsing. All these millions of people who are now publishing blogs on the Web are, to my mind, producing something you might call culture. Of uneven quality, but it's easier to publish now than it used to be. And that to me is not necessarily a disaster. It may be a step forward.

LLOYD: In fact it's easier to preserve information as well. In the past, one of the main problems with culture is that it would disappear, because there was only one copy. When there's only one copy, things get easily destroyed. And yes, maybe because in the United States we don't have as much culture, we're not so worried about losing it. Perhaps worrying about the wholesale copying and monkeying with genetic information might open people's eyes to the danger of copying and monkeying with ordinary cultural information—for instance, violating copyrights. While I am usually for any kind of information manipulation I can think of, it does seem a little strange to try to manipulate human genomes. Of course, the primary way of manipulating genomes in the past, which people have been doing for ages, is by breeding. People are rather squeamish about attempts to manipulate human genomes to create perfect human beings just by breeding. This is an old fear among people—and an old temptation as well. It *is* scary. But anything fun is scary.

CHURCH: Genomics for the most part has been quite open historically—even in profit-making sectors they will publish papers and so forth, and the genome project went as far as to try to publish things within one week of collecting the data. Almost every genome you could possibly want, including some that some people would prefer not to be in open-source, like smallpox, which Craig helped to do, and the 1918 flu virus—all those are available. I think that's a trend.

DYSON: It's unfortunate that smallpox is out there. The world would be a lot safer if that hadn't been published.

VENTER: I can disagree very violently with you on that.

DYSON: Good. That's a minor exception; as a general rule, openness is by far preferable.

VENTER: Even with that, I think I could convince you openness is far more important. There were two states that were funding an incredible amount of secret research—the U.S. and the former Soviet Union—on trying to modify smallpoxes, make them more dangerous, et cetera. So if it were not open-source, those states would be the only ones with access to this information. There would be nothing out there for either tracking it, understanding it, making better vaccines, et cetera, if it was a real threat. And on the synthetic biology side, it's a very, very low threat, because the DNA is not infective. It's a hypothetical threat that people like to use to scare people, but in reality it's not one.

CHURCH: DNA is not infective, but you can make infective viruses with the DNA in the lab?

VENTER: Hypothetically. But nobody's done it yet.

CHURCH: With other pox viruses you can do it, so it's not that hypothetical.

VENTER: There are probably a few thousand pox viruses out and very closely related species that could easily become smallpox. I'll argue for open-source of information. My own genome is on the Internet, but I'm much more selective about whom I share my biological materials with. There's open-source and there's open-source.

SHAPIRO: You did raise an interesting point there, though, because genetic privacy is something often debated—the rights of individuals to genetic privacy. Not to have their genomes known.

VENTER: But that's driven by fear, not by knowledge.

SHAPIRO: But what I'm saying is that genetic privacy may be impossible. Let us say that [Craig] hadn't put his genome on the Internet and wanted it secretive. Say he was running for public office and had some gene for some mental instability and therefore wanted no one to have his genome; yet someone wanted his genome. All I'd need to do is swipe your glass and shake your hand.

VENTER: This is an issue that we could talk about, that George and I have been facing, that's counteractive to what our government is doing. Francis Collins [former director, National Human Genome Research Institute] is setting up databases, for which you have to have retinal scans and fingerprints to get access, and we're publishing our data on the Internet. So, open-source is not a guarantee of any means at all. We hope by making human genetic data available, people will find that it's almost impossible for your scenario, wherein you look at one gene and say, "This person's going to have mental illness." Even the entire genetic code doesn't provide that answer. You have to know the environment; you have to know a lot of other things. Perhaps 50 years from now, we can get much closer to predicting things, but we are not just genetic animals. My dangerous idea is that we're probably far more genetic animals than society is willing to accept. But we're not purely genetic animals, so I don't think it's going to be as predictive as some people think.

SHAPIRO: Well, certain specific things will be predictive—for example, Huntington's disease is due to a repeat of certain letters in DNA.

VENTER: There are some very rare exceptions, yes.

SHAPIRO: You can even tell what onset is likely at what age, by counting the number of repeats that are present.

VENTER: But that's the exception that doesn't make the rule. That's what every geneticist has used as the few early examples of success in genetics of single-gene disorders.

SHAPIRO: But there are cases where individuals themselves didn't want to know whether or not they had inherited the gene for Huntington's disease, or if they did, whether they were going to have a severe form. Yet if some external person wanted to inform himself as to whether that individual did carry the gene, it would almost be impossible to prevent that individual from getting the information. You would practically have to live in seclusion, with all of your clothing, all of your artifacts destroyed on contact.

LLOYD: It's interesting, because in fact the digital nature of genetic information—the fact that it's 7 billion bits that can easily be written into a computer hard drive—makes genetic information much more like the information in computers, and it can be manipulated in that way. Whereas, strangely enough, our mental information, the information that's in our brains, is much less digital in a fashion, and much harder to get hold of.

And in fact it does suggest that—since this information has been digitized, and will continue to be digitized and manipulated and be more available—the question of how to ensure secrecy and privacy for genes is rather similar to the privacy of your iPhone. How privately are you allowed to keep the information in your iPhone? How privately are you allowed to keep the information in your genes? Because it will be available, and it will be possible to get it and to digitize it. So then the question is, Do you need codes for protecting your genetic code? Maybe everybody should be issued their own public-key cryptic system so they and only they can have access to their own genetic code.

CHURCH: We're in a state of change, where we're deciding what's the right thing. For example, consider our faces. Some people keep their faces completely masked; in most situations it's considered antisocial to keep your face completely masked. Like walking into a bank, for example. But [the face is] extraordinarily revealing—it reveals something about your physiology, your current health, your relationship with the person you're talking to, whether you're angry or happy—it's very revealing. And so we've made a conscious decision in society, for the most part, to not keep that private. We might do the same thing for genomes, it could be. Whom are we protecting? It's an open question.

SHAPIRO: Well, we shed cells so easily, unlike faces, that it's almost impossible to keep your genome private if there's someone out there determined to have it.

CHURCH: I agree with you. We'll all become bubble people, living in our little hermetically sealed bubbles so nobody can get in.

LLOYD: Who steals my genome steals trash, right?

J. CRAIG VENTER: Seth's statement about digitization is basically what I've spent the last 15 years of my career doing—digitizing biology. That's what DNA sequencing has been about. I view biology as an analog world that DNA sequencing has taking into the digital world. I'll talk about some of the observations we've made for a few minutes, and then I'll talk about [the fact that] once we can read the genetic code we've now started the phase where we can write it. And how that is going to be the end of Darwinism.

On the reading side, some of you have heard of our *Sorcerer II* expedition of the last few years, where we've been shotgun-sequencing the ocean. We've applied the same tools we developed for sequencing the human genome to the environment, and we could apply it to any environment; we could dig up some soil here, or take water from a pond and discover biology on a scale that people have not even imagined.

The world of microbiology, as we've come to know it, is based on a technology over 100 years old, of seeing what will grow in culture. Only about 0.1 percent of microbiological organisms will grow in the lab with traditional techniques. We decided to go straight to the DNA world, to shotgun-sequence what's there, using simple techniques of filtering seawater into different-size fractions and sequencing everything at once that's in the fractions.

For example, we discovered that almost every microorganism in the upper parts of the ocean has a photoreceptor similar to the ones in our eyes. We knew there were one or two bacterial rhodopsins, but people thought they were rare molecules; it turns out it's probably one of the largest gene families on the planet. It's the same gene family we have in our eyes—our own rhodopsins, our visual pigments. Only instead of just capturing light information, these organisms capture light and convert it into cellular energy—a non-photosynthetic, totally separate mechanism. When we set out to the Sargasso Sea surrounding Bermuda, all the marine microbiologists told us nothing was there, it was a desert, and we'd find only a few organisms. Instead we found tens of thousands of organisms in just a barrel of seawater. And the reason they said we wouldn't find anything is that there are no nutrients there. They said, "No nutrients, therefore no life." It turns out [these organisms] don't need the nutrients, because their energy is derived directly from sunlight.

LLOYD: Do you think perhaps the first use of rhodopsin was to harvest energy?

VENTER: Quite possibly. And then it was adapted for visual pigments, because it was a light-recognition molecule. And the other aspect of it is, there's a wide range of new ones—we have thousands of these now—and lining up the proteins of all these, there's a single amino-acid residue that determines the wavelength of light the receptors see. There's a single base change in the genetic code that determines the amino acid responsible for the wavelength of light seen by the receptor. Changing one base in the genetic code can switch the light seen from, for example, blue to green.

We found, when we went back looking at the distribution of where these different rhodopsin molecules are, they totally segregate based on the color of water. The photoreceptors in the organisms in the deep indigo blue of the Sargasso Sea see blue light. You get into coastal water, where there's a lot of chlorophyll—they see primarily green light. To me, this is classical Darwinian selection. A single base pair determines the switch between blue and green. And whatever wavelength of light, it clearly provides a survival advantage for that environment.

It turns out, just looking at the populations, that this switch between blue and green has happened at least 4 times, back and forth. And so on the one hand, it seems like a classical reformation of Darwinian thinking. On the other hand, under each type of 16S rRNA, we have in fact hundreds to thousands of different cells, different genomes. Are they different species?

These are the types of species and definitions of life I've been lately been devoting much of my professional career to. And it turns out that these diverse organisms share most of the gene content with each other and most of the gene order is conserved. But the sequence variations are as high as 60 percent between cell types. According to classical Darwinism, this should not be the case, and it's not what anybody expected. It should be that one, or a few, species survived and all the others died out.

It turns out that these are ancient lineages, existing in parallel, all virtually identical to each other but not quite identical.

Maybe they just have that equivalent base change—to where one can see the equivalent of blue light and the other can see green light—and we can't see that, because we don't know what all the other molecular switch changes mean. But we find this over and over again, with every microbial species. I'm forced to use the term "species", because there's no better common word, but "species" is a very vague term. It's sort of a group of closely related organisms, much like the human population is a group of closely related organisms. If it's a new species because it has that one base change and it can see blue light instead of green light, that's a phenotypic difference, a survival difference; it's got nothing to do with sexual reproduction, but it's roughly the same order of change of genetic information.

I have come to think of life in much more of a gene-centric view than a genome-centric view, although it kind of oscillates. When we talk about the transplant work, "genome-centric" becomes more important than "gene-centric". From the first third of the *Sorcerer II* expedition we discovered roughly 6 million new genes, which doubled the number in the public databases when we put them in a few months ago, and in 2008 we're likely to double that number again. We're at the tip of the iceberg of what the divergence is on this planet. We are in a linear phase of gene discovery—maybe in a linear phase of unique biological entities, if you call those "species", and I think eventually we can have databases that represent the gene repertoire of our planet.

One question is, Can we extrapolate back from this data set to describe the most recent common ancestor? I don't necessarily buy that there is a single ancestor. It's counterintuitive to me. I think we may have thousands of recent common ancestors, and they're not necessarily so common.

Other things you can throw into the mix: We have organisms that could survive long-term space flight. They can take millions of rads of ionizing radiation, they can be totally desiccated; when they reach an aqueous source, they can repair their genome and start replicating again. Thus you could potentially view evolution as a 6- to 7- to 8-billion-year event, not a 3- to 4-billion-year event, if life can travel around the universe. That adds a lot of dimensionality to things, when we think of life on other planets and galaxies. We exchange roughly 100 kilograms of material annually with Mars. So we're exchanging biological material and biological information. To me it's just a matter of time until life is found on Mars. It's inevitable. It won't tell us whether it originated on Mars or originated on Earth, but there'll be common overlap. We won't know, if we don't know our own planet's genetic repertoire, which we're in the earliest stages of discovering. There are the evolutionary aspects, the origin-of-life aspects, to this, which make it intriguing.

I spent early parts of my career trying to make site-directed mutants of neurotransmitter receptors, where you change a single base-pair and study the change in function of the new altered protein. Now you can just go to a database of our gene sets and find 35,000 variants of a single gene that you know survived and work in nature. It's a different view of the world, and it will change the experimental approaches. But this new data set gives us a wonderful new repertoire, if you view these genes as the design components of the future.

For me, it all started in 1995, when we sequenced the first 2 genomes of living entities. The first was *Haemophilus influenzae;* the second was *Mycoplasma genitalium. Mycoplasma genitalium* had only 550-some-odd genes; *Haemophilus influenzae* had 1,800. This was the first comparative genomics that could be done for living organisms. We started asking simple questions: For example, if one species needed 1,800 genes and the other needed 550, are there species that can get by with fewer? Can you define a minimal genetic operating system for life? Could we define life at a genetic level? Obviously extremely naïve questions, but the view of biochemistry and genomics by the scientific community was limited as well. For example, when we published the *Haemophilus influenzae* genome, a well-known biochemist at Stanford University said we had obviously assembled it wrong, because it didn't have a complete TCA [tricarboxylic acid, or citric acid] cycle, and everybody knew that every organism had a complete glycolytic pathway and a complete TCA cycle, and *Haemophilus* had only half of one.

SHAPIRO: Therefore it's not an organism?

VENTER: No, therefore they assumed we'd made a mistake in the sequencing and the assembly. Now we see every repertoire under the sun; for example, the third organism we sequenced was the first *Archaea* we did with Carl Woese. It was *Methanocaldococcus jannaschii*, which has neither a TCA cycle nor glycolysis. It makes all its cellular energy by methanogenesis, going from CO2 to methane, using hydrogen as its energy source. CO2 is its carbon source for all the carbon in the cell.

What Freeman was talking about when he said you can separate metabolism from replication is certainly true, and we have at least 20 different modules we could plug in for metabolism. There's not a universal genetic code that will be an operating system; it can be a choice of methane production or of glucose metabolism or anaerobic metabolism, oxidative metabolism, or some other approach. So the naïve assumption that we could even define an operating system clearly went out the window.

We first tried to knock out genes in *Mycoplasma genitalium* to see if we could get to a minimal genome. We could knock out only one gene at a time. We could knock out about 100, but one at a time, but that didn't tell us whether, if you knocked out all 100, you could get a living cell. But we also learned other things: for example, that essentiality for life is very much a relative term. We can have a gene that's absolutely essential for life in one circumstance but not another. The simplest example I give is that there are 2 genes in *Mycoplasma genitalium* for sugar transport—one for fructose and one for glucose. If you have both sugars in the media and you knock out either of the transport genes, you would say they're nonessential genes, since the cell lives. If you have only glucose as your sugar in the media and you knock out the glucose transporter, the cell dies, and you say, "Ah, that's an essential gene!" All these things are definitional, based on two things: the genetic code and the environment.

Right now, we're all focused on the genetic code because it's something we can define, and the environment is so many orders of magnitude more complex to define. But we're having this trouble with a single cell with a few 100 genes. We, as humans, have 100 trillion cells with 23,000 or so genes and an infinite number of combinations, so defining our environment is going to be a lot more complicated than doing that for a single cell. We decided the only way to answer these questions was to make a synthetic chromosome, to understand minimal cellular life.

We set out with some simplistic experiments in 1995 to make the Phi X 174 viral genome. My colleagues in this—and I have an incredible set of colleagues—included Ham [Hamilton] Smith, who discovered restriction endonucleases, which led to his Nobel Prize in 1978, and Clyde Hutchison, who was in [Frederick] Sanger's lab when he sequenced Phi X 174 and was one of the inventors of site-directed mutagenesis. We thought we'd just synthesize a set of overlapping small oligonucleotides, anneal them together with a replicating enzyme, and we'd get the complete genome. We decided to start with Phi X 174 because of its historic value and because there are very few base pairs in it that can be varied and still get a functional virus. We figured it would be a real test, because you had to have it accurately synthesized to get a functional virus from it. And even though we had selection by infectivity, which gives us a million- to 10-million-fold selection, even though we got full-length genome molecules, none of them were viable.

And I use the term "viable" loosely with viruses; it's not clear what a viable virus is, but it's one that can infect a cell and start the replication machinery going. At the same time, we decided that if the only approach forward was to synthesize a genome and then start modifying it, we were going to be creating new biological entities, or species, that hadn't existed before.

It's true that if you knock out a gene—if one base-pair difference is sufficient for something new—this is nothing new in molecular biology, except if we try to do this deliberately and then try and design things from it. So we asked Art Caplan at the University of Pennsylvania to spend time with his [medical ethics] group looking at this, and his team brought in every major religion to see if it was OK to synthesize life in the laboratory. After a year-and-a-half study, basically, none of the religious groups objected to it, because they couldn't find anything in their scriptures that said it shouldn't be done. In fact, they mostly came up with the opposite of this notion of "playing God" that the lay press talks about all the time with regard to this and other related work. Every major religion basically said that part of their dictate to humanity was that you were supposed to try and use knowledge to benefit humanity. The report they published in *Science* in 1999 noted that what we were doing, and the way we were approaching it, was reasonable and that we should proceed. The only caution was about biological terrorists using our techniques to try and make biological weapons.

We began with that as a basis. Then the entire project was postponed, because I had the opportunity to sequence the human genome. We gave up on synthesizing new life-forms for a while.

LLOYD: That's interesting. So you actually came around to sequencing the human genome from the perspective of trying to construct an artificial organism.

VENTER: They were parallel tracks. One didn't lead to the other. I wouldn't want you to make that intellectual leap.

LLOYD: But you were trying to synthesize life, then stopped doing it in order to sequence.

VENTER: Right. And so we've spent basically since 2002 starting in full form to find ways to synthesize genomes. We started back with Phi X 174, and Ham and Clyde came up with some new ways of error-correction and improving synthesis. DNA synthesis is a degenerative process, wherein the longer you make the molecules, the more errors are in them. Trying to just synthesize off of chemical synthesizers and get really accurate molecules is currently not possible.

But it was an exciting leap for us when we actually made this chromosome, injected it into *E. coli*, and the next thing we knew, *E. coli* started using that synthetic genome to start producing the Phi X 174 viral particles, which then started killing the *E.coli.* Clearly, this human-made piece of chemical DNA software was now building its own hardware. One of the exciting parts of synthetic biology and synthetic genomics is that this is possible.

There are two real components or questions to this, and there's even more when you think of the implications. One is, Can you make these large molecules? And the answer is absolutely yes, we can make whole chromosome-size macromolecules of DNA. While we were thinking about the synthesis, we were thinking, How do we boot up the chromosome in a cell? In the process, we initially thought, "Well, you'd like a ghost cell that just has the ribosomes and other cytoplasmic components in it but is devoid of its chromosome," and we tried numerous ways to get rid of a chromosome in a bacterial cell. Then Ham Smith came up with the notion of, "Well maybe we don't really need to get rid of it; we'll just put the new chromosome into a cell and when they segregate maybe the chemically-made chromosome will go into one daughter cell and the other one will go another way." Actually it's much more complex than that.

A short while ago, we published a paper in *Science* on genome transportation ["Genome Transplantation in Bacteria: Changing One Species to Another, *Science*, Aug. 3, 2007], where we took a purified chromosome from one species, made sure it was totally devoid of any protein, and put that chromosome into another bacterial species. It's the ultimate identity theft, because the new chromosome we put in completely took over the cell, and the cell converted completely into the cell dictated by the new chromosome. We put an antibiotic selectable marker gene into the transplanted chromosome so we could select for those cells with the new chromosome. The story on how the selection happens and why one chromosome survives is actually much more interesting and deals with an important part of evolution, but needless to say the new chromosome dictated everything. All the proteins changed over to that. The phenotype of the cell, everything changed, converted from the old species into the new species.

DYSON: How many generations did that take?

VENTER: It's not clear. It could have happened in the first couple of generations. Until you get enough cells that you can see and do biochemistry on, they've gone through dozens of generations. It would be nice to do stop-flow experiments and see what happened in those initial phases. We tried to take an EM-micrograph to see if there were hybrid cells that had both sets of chromosomes in them, but we did not find any evidence for hybrid cells. You can see why restriction enzymes were so important for cellular evolution, because speciation could have problems with DNA uptake if whoever had the dominant genome could just immediately take over your species and transform you into them. It's true identity theft at the molecular level.

LLOYD: And you wonder why people are worried when you describe it like that. I feel that it's happening to me right now!

SHAPIRO: How genetically different were the two cells?

VENTER: They were roughly the equivalent of man versus mouse. They were closely related mycoplasmas, and it turns out the restriction barriers are major barriers. It became clear to me for the first time how important restriction enzymes were for evolution, because if foreign DNA can go in and take over the cell, obviously that's what you want protection against. It's their equivalent of an immune system. To do genome transplants on a reasonable scale, we have to overcome the restriction barriers. In the case that worked, the chromosome we transplanted had a restriction enzyme that we think chewed up the chromosome that was in the cell. The one in the cell did not have a restriction enzyme, so there was no restriction barrier.

SASSELOV: Is your gut feeling that there is no need for a hybrid generation to develop, that you jump entirely from one to the other?

VENTER: I don't know.

SASSELOV: But do you have a gut feeling for it?

VENTER: Gene expression and new protein synthesis can happen pretty quickly. If we're transplanting a new chromosome and it immediately starts to get transcribed, events could happen quite rapidly. Turning over the membrane and all its content could perhaps take a couple of cell divisions and generations. After several generations, we ran 2D gels and some protein sequencing; every protein there was one dictated by the new chromosome.

The genome-transplant experiment was the key one for synthetic life. But definitions are important here. None of us is talking about creating life from scratch, because that's not what's happening. We're taking two approaches: We're taking the genome-transplant approach and we have a team working on trying to isolate every protein from the cell to see if we can put the proteins together, with the chromosome, with some lipids, and get spontaneous cell and life formation.

LLOYD: You're saying it's much less like open-source software and more like Microsoft, whose software actively resists operating with exterior software.

VENTER: Absolutely. You could not have speciation without it.

These things get down to basic definitions of life. The lay press likes to talk about creating life from scratch. But while we can create and develop new species, we're not creating life from scratch. We talked about the ribosome; we tried to make synthetic ribosomes, starting with the genetic code and building them. The ribosome is such an incredibly beautiful, complex entity. You can make synthetic ribosomes, but they don't function totally yet. Nobody knows how to get ones that can actually do protein synthesis. But starting with an intact ribosome is cheating anyway, right? That is, not building life from scratch but relying on billions of years of evolution.

When starting with an existing protein-synthesis machinery, we can create new life-forms, we can create a synthetic chromosome we can now do transplants of and develop new species with unique properties. So we can create human-made species, but we're not really creating life from scratch. You can boot up a system, but right now all life derives from other living entities. What we're doing is really no different, because we're just putting a new operating system into a living cell.

If George [Church] or anybody else doing this can take even

the raw protein components and lipids and boot up a cell and get it activated and go from the chemical molecules to a living cell, that's a big conceptual barrier that remains to be passed. But even that's cheating. If you look at the Urey-Miller experiments, on the chemicals that can be made in certain environments, and start with basic amino acids and nucleic acids and all of a sudden you get life out of that—that would be creating life from scratch. I don't think we're creating life. We're coming up with new modified life-forms, and we should be able to go from the digital world right to the analog world in the computer, and we have a team working on a program to do that, designing a species in the computer.

It's only a short time away from doing that just to have systems crank out synthetic chromosomes. In fact, I've talked to various funders about trying to design a robot that could build a million chromosomes a day. Because then we can have a new field, which I call combinatorial genomics. We don't know answers to seemingly simple questions such as, Is gene order important? If you just scramble the order of the same genes, does life work the same way? We know with operating systems and operons that gene order is important, but in the genome as a whole it is perhaps not. That was one of the biggest surprises with the human genome—multi-subunit proteins such as the nicotinic acetylcholine receptor that has 5 sub-units which were all in different chromosomes. The assumption was that they'd all be together, in order, on one chromosome. Maybe as long as all the parts are there, that's all that matters—we just need the gene set and the gene repertoire. There are so many different directions we can go with this conversation, maybe I should just stop here.

BROCKMAN: Where is it all going?

VENTER: I disagree with this recreational use. Who cares whether you have green fluorescent protein in fish or something? Hopefully, these are fads that will go away quickly. Maybe they're important for inducing biological concepts, but there are far more pressing issues.

To me, the biggest issue—that's why we put most effort there—is what we're doing to our own environment and atmosphere by taking billions of tons of oil and coal, burning it, and having the CO2 go into the atmosphere. It's a big experiment we're doing with our planet which hasn't happened during the existence of human life. It's a dangerous experiment to do. We can only estimate outcomes from it. But we have to have some potential solutions. While everybody's looking to physics for the solutions, I've been arguing that biology could play a major role, if not the ultimate role, in the solution. And that's why we started Synthetic Genomics, the company, to try and design genomes to make new fuels.

DYSON: We were doing that at Oak Ridge 30 years ago.

VENTER: People have been looking at biology, but they were looking for naturally occurring organisms to do this. Right? That worked well during World War II.

DYSON: It was clear that biology would be the right way to do it. Even 30 years ago.

VENTER: There has been an effort within some DOE labs to find and use natural organisms to produce hydrogen or other potential fuels, but the efforts have essentially gone nowhere, due to the scale of the problem, and the efforts have been limited. It has been a battle in the Department of Energy, as to the importance of biology. Many in Congress have the naïve idea that it should only be done at the NIH [National Institutes of Health].

LLOYD: If we destroy our idea-based culture by raising the ocean levels by 100 meters, we can just return to the good old days of Darwinian natural selection. Go back to stage 5, right?

VENTER: The choices are doing something to change the environment or engineering the human genetic code to be able to survive in different environments. When we can actually design and build millions of new organisms a day, single-cell organisms—the first single-cell organisms are months away; the first synthetic eukaryotic cells are less than 5 years away, and multicellular systems are not orders of magnitude more complex to do. In fact, in some respects they're easier to do, because the ocean is a multicellular genome system, just different cells provide different components—the cells can be associated loosely in the environment, but there are only a small number of cells that actually fix nitrogen that provide that for the whole pool. It's a cooperative environment, and having the cells get together to do it is not much more complicated than engineering some good stem cells, if we want to do that.

SASSELOV: Craig, I wanted to connect what you're saying with what Freeman was saying about the Darwinian era. To me, his idea is not so much about the end of a process but the beginning of a new phenomenology, from the big-picture point of view. Don't you feel that creating those species, or whatever you may want to call them—synthetic life, even if it's not creating life from scratch—basically starts a new phenomenology in the universe? Because you have a complex chemistry that reached the stage at which it could actually change and produce viable complex chemistry that would continue—even without its own existence. In other words, if we do not continue as a species, and our technological civilization comes to an end, those species will continue to exist on this planet and potentially could go to other places?

VENTER: Yes, it's an important conceptual change.

LLOYD: Human beings, you could even say, could identify the beginning of modern human culture with the ability to modify grains genetically so they could be grown in large quantities, and people could go from being hunter-gatherers to being farmers. And certainly people have been genetically modifying the world around them for tens of thousands of years, in a variety of different ways, starting merely by first picking the variety of grains that have things that are easier to separate from their husks. The recent discovery of the precursor of corn, or maize, was a big surprise, because the precursor looks nothing whatsoever like the corn we eat, and over the thousands of years the people, merely by selecting out the varieties they wanted, made corn as we know it.

VENTER: We've been doing blind genetic experimentation by mixing whole genomes together for a long time. It's amazing how little concern there's been over the agricultural practices for the last millennium, right? Just blindly mixing any species together—if they will mix together, it gets done. And if you do it with intelligent forethought, it's more dangerous, somehow.

LLOYD: I guess one of the worst consequences of hybridization was that the species that were introduced were so successful that we got rid of a huge amount of biodiversity by replacing what used to be a much larger set of different kinds of corn with just a couple of varieties, which are then much more vulnerable to various blights.

BROCKMAN: What drives the decision as to what projects you pursue at the [J. Craig Venter] Institute? Available funding? Areas that you and your colleagues picked out? What you can get away with? Are you looking for another definition of life? And finally, what's your fantasy—if you had all the money you wanted and needed and you weren't going to get hassled by the government, or the press, what would you do?

VENTER: I'm one of the few scientists actually in the situation to live most of his fantasies every day. My institute budget varies between $80 million and $100 million a year in funding, the majority of which comes from federal sources, but an increasing percentage comes from the wealth that's been generated in this country. The Gordon and Betty Moore Foundation—Gordon was the founder of Intel—is an example of people who have made tremendous private wealth, and put it back into science and to the benefit of society. It's a unique thing in this country—I don't see it anywhere else in the world. More and more of our research funds come from those sources. In addition, out of founders' stock from Celera and Human Genome Sciences, I created my own endowment that can fund new ideas when they occur, not a year or two later. I've found that's the key in creative science, at least in my case: both having resources and the ability to do experiments when I think of the ideas.

LLOYD: I want you to stop describing this, because I just might have to leave the table if you continue. It's too depressing for me.

VENTER: We could have an entire session on just how dismal new science funding is in this country. We celebrate the breakthroughs, but to me they happen at 0.001 of the rate that they should be happening. But I've had the privileged situation by creating the environment I want, so I and my colleagues do things driven totally by our intellectual ideas. And we spend our seed money on it, and then we try to find other sources to fund them to the next stage.

BROCKMAN: What are the ideas that are too dangerous to pursue, that you want to pursue?

VENTER: There isn't anything now within technical capability that's worth doing that is too dangerous to pursue. Our knowledge of the human genome is so primitive that to start engineering it is just stupid. Hopefully, in 50 or 100 years, our knowledge will be sufficient that we could do that intelligently. In the long run, genetic manipulation of humans is not only inevitable, it's probably a very good idea.

GEORGE CHURCH: We've heard a little bit about the ancient past of biology, and possible futures, and I'd like to frame what I'm talking about in terms of 4 subjects that elaborate on that. In terms of past and future, what have we learned from the past, how does that help us design the future, what would we like to do in the future, how do we know what we should be doing? This sounds like a moral or ethical issue, but it's a practical one too.

One of the things we've learned from the past is that diversity and dispersion are good. How do we inject that into a technological context? That leads to the second topic, which is, if we have some idea what direction we want to go in, what sort of useful constructions we'd like to make—say, with biology—what would those useful constructs be? By "useful" we might mean the benefits outweigh the costs and the risks. Humans as a species have trouble estimating the long tails of some of the risks, which have big and unintended consequences. So that's utility.

Many of the people here worry about what life is, but maybe in a slightly more general way— not just ribosomes but inorganic life. Would we know it if we saw it? It's important as we discover other worlds, as we start creating more complicated robots, and so forth, to know where we draw the line. I think that's interesting.

And then, finally, the kind of life we're particularly enamored of—partly because of egocentricity but also for philosophical reasons—is intelligent life. But how do we talk about that?

As a scientific discipline, many people have casually dismissed intelligent design without carefully defining what they mean by intelligence or what they mean by design. Science and math have long histories of proving things and not just accepting intuition—Fermat's last theorem was not proven until it was proven. And I think we're in a similar space with intelligent design. What Freeman suggests is that we're moving into a phase that's different not only in that it's like Web 2.0, where we're all sharing all of our parts like we used to, but, more fundamentally, we're moving into intelligent design big-time and we need to understand what that means and what we should be designing.

In terms of utility: People might have huge disagreements, even within a religion, as to the right thing to do. You might say,

"Thou shalt not kill"; the same person a few days later might say you must kill—you must kill a lot of people. What might we all agree on? Well, we might agree that it's not a good idea to wipe out the entire intelligent species in the universe. Even if you believe in the afterlife. Then you might say, "Well, we shouldn't kill off the afterlife." There's some basic thing we like, and it has to do with complexity and what we mean by intelligence. We'd like to preserve that somehow, and I apologize for that being fairly philosophical, but when we start to construct life, what is it we're trying to do? We're trying to make things that are more complex. But it's not just complex. You take a rock: essentially quite complex. If you take a leaf, it's complex in a different way. Take that leaf and then rearrange all its atoms so it's just a bunch of salts in a rock—ammonium carbonate and silicon dioxide and potassium phosphate and sodium sulfate. It's the same atoms but in a form that a mineralogist would recognize. The mineral is still complex; if you wanted to transmit over the Internet the structure of that mineral, it would take a lot of bits.

Both [Claude] Shannon's theory and chemical entropy would say that's a very complicated thing. But what we mean by a living complexity is more like you've taken something rare, like that mineral, made almost an exact copy of it, and that's replicated complexity. This unlikely object isn't really any more unlikely than another rock, because they're kind of random—their compositional nature is known. But if you made an exact copy of that rock, or nearly an exact one, that would be interesting. That would be indicative that some sort of living process, some living thing, was involved. It could have been some 3-D photocopier, but that 3-D photocopier was probably made by an intelligent being. It could be that the rock had the ability to replicate.

That's what we mean when we're talking about basic life. And that's what we're trying to get at when we're doing synthetic biology; we're trying to increase diversity, increase replicated complexity, and maintain our ability to continue to do that for many many years, and we don't want to endanger that by doing something that's too risky.

LLOYD: Are you implying that there's a virtue in increased complexity somehow?

CHURCH: I'm trying to make that argument. There might be a virtue in carefully contemplating not just short-term diversity but longish-term—to the extent that we can calculate that, which we can't right now. But it's desirable to be able to calculate that as well as we can, and I'll just take a leap at defining intelligence, too, while we're in dangerous territories here for me.

There's analytic intelligence and synthetic. And I would argue that life is sort of this replicated complexity, or mutual information, where, given the molecules in this leaf, we can predict the arrangement of structures in the other leaf. In other words, we know a lot about this thing—even within the leaf there's replicated complexity that's somewhat predictable—and so that mapping, that mutual information, is predictive of life in general. But the mutual information is something where one structure will reflect the structure of something at a distance—especially if you can reflect something distant in time without actually causing it. Intelligence is anticipating things in the future without causing them. That would be analytic intelligence. Again, it's replicated complexity—or mutual information, even better—where there's a relationship between the two but you anticipate. That's analytic.

Synthetic's harder, because if you synthesize something, you've used your analytic intelligence to make a plan and then make a replicated complexity of some sort at a distance, but you've done it. There's a cause and effect. I'm still struggling with this, but I think synthetic intelligence would be something that in one way or another would enhance analytic intelligence—our ability to predict what's going to happen in the future. So we synthesize something that will increase our ability to, say, survive as a species, to get off the planet—because we know an asteroid is coming. Various things we would recognize as long-term intelligent behavior. And it might be inorganic life that does that, in the form of computers, or maybe some hybrid.

I'll come back down to earth a little more and talk about synthetic biology, but part of getting at intelligence is assaying what kind of intelligence is already on the planet. And what's really remarkable is, we're still far ahead of our computers, in that we have 6.5 billion geniuses on the planet. Some of them are undereducated, but there really aren't any computers comparable to people—to brains—so far. That may not last forever, but it's certainly true now. We need to assess that diversity and not cure it; as we get better at personalized medicine, the goal is not to cure our diversity but to enable it, to make it so we can all enjoy our lives and contribute to all the other things we've been talking about today.

Personal genomics is getting into the analytic phase of figuring out what we all are capable of; synthetic biology is still very primitive. We're interested in useful things, like making fuels. Craig already set this up: If we're going to burn carbon, of which there are vast amounts in the Earth, we might want to have some way of recovering that carbon and maybe burning it again. If we run out of petroleum—it's not only useful as a fuel for cars, trucks, and planes but it's also the source of a lot of our constructive plastics. We need a replacement for petroleum, at least in the short term. One of the companies I've founded is called LS9, in California, and they're making synthetic petroleum, as an example of what you can do with synthetic biology. Hydrocarbons compatible with current engines—cars, diesels, and jets—rather than requiring a new infrastructural change, and that's basically metabolic engineering and exchanging DNA from many organisms, just as Freeman said—not just one gene at a time but whole systems, whole metabolic systems, and using what we know to expedite that.

Another company I've been involved in is called Codon Devices, and it makes the DNA for companies like LS9. They have a capacity of about 2 million base pairs of very highly polished DNA per month, which goes out to many biotech pharmaceutical companies. It's not making whole genomes, or whole chromosomes, like Craig talked about, but it can make several such genomes per month. And it's important to have this, and to have it regulated. Codon Devices joined with about a dozen other companies to increase their ability to monitor DNA synthesis worldwide. I started this with a white paper many years ago, tried to bring government attention to this, and it became clear that the government isn't going to act unless it's sure that industry isn't going to be hurt. It isn't always a formula that will work—but if industry can get its act together and do it voluntarily, then governments can point to it and say, "Oh, that's already working." They're sharing resources, so the cost of software is distributed, the cost of monitoring local regulations is distributed, and then the government can say, "OK, we'll make the law." At which point, internationally, they can say, "Many governments are acting; we can make international law." I hope that's the way these things work out.

That's synthetic biology—making biofuels, helping improve

stem-cell biology; you don't necessarily have to start with a germline if you can change people's soma. That may be less risky; it may allow you to do more rapid prototyping and alleviate people's concerns. The germline is a special set of concerns, and there's quite a bit you can do—there's actually more you can do, in a way—because predicting what somebody's going to be like from a fertilized egg is hard, but predicting what they're going to be like once they're thirty years old is quite a bit easier. It's not really prediction, it's just observation. It may be harder to fix, but at least you'll develop tools and you'll be more cautious.

There's a great deal of progress there; now we can establish more pluripotent cells from, say, mammalian skin cells, which will enable synthetic biology to move in that direction. And there's even some projects we're working on as part of an NSF-funded project with Berkeley, UCSF, and MIT, to engineer bacterial cells compatible with mammalian immune systems so they can motor around inside your bloodstream and do sensing and actuating. For example, they'll home in on tumors. They'll sense their presence; there will be a 1,000 times higher concentration near a tumor—all the parts of this are working, but the whole is not working—and they'll sense they're there, they'll invade the cell by expressing an invasive protein, and then inside the cell they'll make a drug that will destroy the tumor cell. That kind of capability, of working well within a mammalian immune system, can range from using your own cells that are perfectly immune-compatible to using these engineered bacterial cells.

There are many other things I could say, but they could be more easily said in the context of questioning. Hopefully I've brought up enough provocative points for you to ask interesting questions.

BROCKMAN: How is your work different from Craig's?

CHURCH: Well, he's much more productive.

VENTER: I use George's techniques.

CHURCH: There, isn't that sweet of him? We develop technology, for the most part. Usually we try to enable other groups to do production—I'll start a company or I'll work with some genome centers to get our sequencing or synthesis technologies to work. From a synthetic standpoint, the major difference is that Craig is a little more interested in making a synthetic genome from scratch; we're mainly interested in making variations on genomes, although I'm sure both of us would be comfortable doing the other as well.

He mentioned making combinatorial a million chromosomes. Well, for instance, we do that; we do lab evolution, and we make millions of chromosomes that go into competition with one another. You either do it by recombining every base pair, spontaneously, in which case you can get only one mutation at a time—we've gotten up to 3 or 4 mutations serially in that way—or we can do site-directed mutagenesis, and we have a new automated method of doing that. Or we can get a series of 23 mutations in 9 days—one at a time, for up to 3 hundreds of days.

One interesting thing I didn't mention and Craig was kind of getting at—whether we're building things up from atoms—is we are trying to make a mirror-image biological world, starting with DNA polymerase, and that does, in a certain sense, require starting at a much more fundamental level than synthetic biologists usually do. Synthetic biologists in our classes will PCR genes up by the cells—that's dependent on life. Craig and I will synthesize genes from nucleotides—that's basically just dependent on knowledge and a few chiral centers, but if you actually flip the chirality, now you're really closer to dealing with atoms.

LLOYD: Do you do this for safety reasons?

CHURCH: That's right, Seth: For every one of these things, we should ask why we're doing this. Why we're doing molecules is obvious. Why we're doing stem cells and pharmaceuticals is obvious. We're changing the genetic code in the normal chirality for safety reasons, and to extend the number of amino acids. In each of these cases, you have to disable it in some way, because changing the chirality makes it incompatible with the rest of the world, but that can make it more of a threat or less of a threat, depending on what other things you do. It could be more of a threat, because now not only is it resistant to phages, it's resistant to enzymes, like proteases and ribonucleases, and at least existing antibodies. Now, if you put in a mirror-image cell, you would get new antibodies, and that's not a problem.

The other reason we're interested in it—aside from safety, which is always something we're interested in and a major theme in synthetic biology and Codon Devices—is, let's say you can evolve DNA and RNA molecules that will bind to your favorite thing. In a certain sense, this is morphogenesis from scratch: You make a completely random selection of polynucleotides and you can get some that will bind to your favorite surface or molecule. In a way, you didn't define that surface; you found it, in a random collection. That's getting close to what we'd like to get. When you use those in a practical setting, they get degraded by biological fluids. But if you made the mirror-image form of it—and since you're starting from scratch, you don't have any preconceived rules, you're just evolving by selection of binding—you can start with a mirror-image nucleic-acid set, a library containing trillions of molecules, and you'll find something that binds to your favorite molecule and is resistant to the enzymes. That's one motivation for the first thing we're making, which is DNA polymerase. We want to be able to mirror-image PCR—polymerase chain reaction—where you can amplify DNA with a mirror-image polymerase. A postdoc has gone through a prototype polymerase for a medium size—353 amino acids long—and he's made all but 4 of the peptide bonds now. So we're close to getting that first polymerase.

The mirror-image nucleotide part is fairly simple, because you can use the same machines to make mirror-image DNA. The peptide-synthesis machines are much more primitive than the DNA-synthesis machines, so a lot of this is done currently by hand. But the goal then, so we can make mirror-image DNA, is to make mirror-image proteins, and there we have to make all of the ribosome from scratch; this is something we think is useful. And that's about 25 times more bonds to be made than just making the DNA polymerase. But, as Craig will attest, scaling up by a factor of 25 is not that big a deal. In the genome project, we went basically 100,000-fold scale-up from where we were at the time we started. And now we're talking about doing many, many genomes. I think we'll be able to make a mirror-image DNA polymerase and ribosome, in which case you can start programming it straight from the computer. Once you have it all, you can start making mirror-image proteins.

VENTER: There's the big assumption that the mirror image of all these things will have the same activity.

CHURCH: Will have the mirror activity.

VENTER: Will have the mirror activity on the other chiral molecules. Is there any evidence for that?

CHURCH: There is, a little bit. I wish there weren't; then we could get all the glory. But I think the HIV protease has been made in mirror form and has been shown to be inhibited by things that are in the mirror form. Very small things have been made. And crystallography has shown that the mirror-image polymers—they make up the mirror-image monomer—the mirror-image polymer is flipped. Almost every time I mention this, there's a subset of people who feel that it couldn't be otherwise and there's a subset that says, "Prove it." I'm happy either way.

SASSELOV: That brings me to a question here. When you talk about synthetic biology, you feel that in the next few years things will evolve to what Craig called "life from scratch." Do you think there will be a clear watershed along the way, or will it get there incrementally?

CHURCH: Well, almost certainly it's going to be incremental. There'll be many milestones. Certainly Craig's article in *Science*, "Genome Transplantation in Bacteria," was a milestone. When he does it with synthetics, that will be one. Another will occur if we manage to get synthetic ribosomes in a chiral form. There'll be many milestones, but every one of them you can trace to something incrementally similar in the recent past.

SASSELOV: But you feel there is not one big gap that needs to be crossed and then you're there?

VENTER: The gap of taking inanimate objects and getting life from them is a hurdle that, when crossed—it's inevitable that it will be crossed, and relatively soon, somewhere—intellectually it's not a gap, but until it's done it's a big conceptual gap.

CHURCH: That's almost philosophical—whether you got the atom from carbon dioxide or from ribose—but I see most of the gaps as practical ones. This field will go faster the more useful it is, and people will resist it less the higher the benefit-to-cost-and-risk ratio. Most people accept evolution; even creationists accept microevolution. If we start getting macroevolution in the lab, then they'll accept the macroevolution to whatever extent it's useful and obvious. If it's not demonstrated in the lab, then you might reasonably say, "I don't care" or "Prove it." The *scientists* should be saying, "Prove it. Do it in the lab." Now, some things, you argue, can't be, but macroevolution is something that might be possible—we're certainly doing microevolution big-time. A lot of companies now depend on pretty amazing changes in the structure—you could argue, though, that they all have intelligent design somewhere in the process. But I think the less intelligent the design and the more macro the evolution, the more people will accept it.

BROCKMAN: Can you talk about biofab labs and their self-replicating nature—in terms of the discussion about fooling around with the human genome and playing God?

CHURCH: You're certainly not creating a universe, you're constructing things. Fab Labs are very much a continuation of all the other engineering disciplines—civil engineering, electrical engineering, mechanical engineering, chemical engineering. Ironically, genetic engineering was really not what most engineers would recognize as an engineering discipline when the term was coined. They do recognize, or they are part of, the revolution that's now finally making it an engineering discipline, with interchangeable parts, hierarchical design, interoperable systems, specification sheets, that kind of stuff. Stuff that only an engineer could love.

BROCKMAN: What's the difference between Neil Gershenfeld's Fab Lab and your biofab labs?

CHURCH: I was just at the annual meeting of the Fab Labs of the world which Neil Gershenfeld organized in Chicago, and I did make that comparison. On the plus side, the current generation of Fab Labs interoperate well with computers, while biology basically doesn't—with the exception of what Craig and I have been talking about.

Take native biology. It's very hard to stick a WiFi onto a corn plant, while in Fab Lab it's all about that, interacting. On the positive side, despite some efforts, there's no inorganic or nonlife technology, despite sophisticated Fab Labs, capable of making itself. A Fab Lab can't make itself without a huge amount of human intervention, which is something the most elementary bacterium can do. And even with human intervention, there's not some compact Fab Lab; it's something that's spread out over continents—there'll be one place that makes the integrated circuits, another place that makes the nice steel bars you use, another place that turns petroleum into plastic, and so forth. It's not an obsession with them, but something that Fab Labs toy with is making a compact desktop device that could make copies of itself. They—to some extent like us—want to have an open-source environment. They already do this in the Fab Labs; they'll send over the Internet plans for making a chair, or a house, and they'll make it in another country, without actually physically transferring a person or a device. That's very exciting and something we share in common.

QUESTION FROM THE PRESS: It sounds like the kinds of tools you're talking about can also be useful for doing some sort of hard experimentation, or at least testing different theories of the origin of life. Not starting from the bottom up. You can at least approach it from the top down and sort of pick apart the different models. Is that a direction you're involved in? Or are the people who are using your tools doing that?

CHURCH: I'm a little more interested in the future than the past—but I don't dismiss it either. For example, at the top of Freeman's wish list was ribosome archaeology. And Dimitar asked, "Is there a milestone we think is significant?" The ribosome, looking both at the past and at the future, is a very significant structure—it's the most complicated thing that's present in all organisms. And it's recognizable; it's highly conserved. So the question is, How did that thing come to be? And if I were to be an intelligent-design defender, that's what I would focus on; How did the ribosome come to be?

The only way we're going to become good scientists and prove that it could come into being spontaneously is to develop a much better *in-vitro* system, where you can make smaller versions of the ribosome that still work and make all kinds of variations on it to do useful things that are wildly different, and so forth, and get real familiarity with this complicated machine. Because it does a really great thing: It does this mutual-information trick. Not from changing something kind of trivial—from DNA to RNA; that's really easy. It can change, from DNA, 3 nucleotides into 1 amino acid. That's really marvelous. We need to understand that better.

VENTER: And you can't have life without it [the ribosome].

CHURCH: Definitely. It's common to all life. We need to understand that, and the way we're going to fund it—there's not that much funding for prebiotic science, but if there's a lot of funding for understanding the ribosome, inevitably it will enable studies of it in the archaeological and ancient-biology sense.

VENTER: But using these tools, it's my hope we can do something similar to what you suggest. We can extrapolate back, once we have the database of Planet Earth genes, to what might have been a precursor species, and then we should be able to build that in the lab and see if it's viable, and then start to do component mixtures to see if you can spontaneously generate such things.

CHURCH: But isn't it the case that, if we take all the life-forms we have so far, isn't the minimum for the ribosome about 53 proteins and 3 polynucleotides? And hasn't that already reached a plateau, where adding more genomes doesn't reduce that number of proteins?

VENTER: Below ribosomes, yes: You certainly can't get below that. But you have to have self-replication.

CHURCH: But that's what we need to do. Otherwise they'll call it irreducible complexity. If you say you can't get below a ribosome, we're in trouble, right? We have to find a ribosome that can do its trick with fewer than 53 proteins.

VENTER: In the RNA world, you didn't need ribosomes.

CHURCH: But we need to construct that. Nobody has constructed a ribosome that works well without proteins.

VENTER: Yes.

SHAPIRO: I can only suggest that a ribosome forming spontaneously has about the same probability as an eye forming spontaneously.

CHURCH: It won't form spontaneously; we'll do it bit by bit.

SHAPIRO: Both are obviously products of long evolution of preexisting life, through the process of trial and error.

CHURCH: But none of us has re-created any.

SHAPIRO: There must have been much more primitive ways of putting together catalysts.

CHURCH: But prove it.

VENTER: You need to improve DNA synthesis a little bit, so that it's 3 or 4 orders of magnitude faster. Then you can make a seemingly infinite pool of nucleotides and start to get— To me, the key thing about Darwinian evolution is selection. Biology is 100-percent dependent on selection. No matter what we do in synthetic biology, synthetic genomes, we're doing selection. It's just not *natural* selection anymore. It's intelligently designed selection, so it's a unique subset. But selection is always part of it. We're not that far away from being able to do these experiments. It's very hard to do now, because nobody will spend the money to make all these different related molecules to see if we can get spontaneous ribosome formation, but within a decade it will be doable.

LLOYD: I would be a little bit worried. If I look at Freeman's two steps that precede formation of ribosomes, ribosome is step 3, with collaboration and intervention of the ribosome, and you have these two steps prior to that. Before is the parasitic stage and use of ATP, and then prior to that just the garbage bags on their own. There could have been a lot of events of natural selection going on to get to the stage—it could be a very, very, very long process, with Avogadro's number of events. There are not enough graduate-student lifetimes in the world, even with lots of private money invested, to try to explore all those. Even if life just happened here on Earth, it's something that happened globally and it went on for quite a long time. I'm saying this in a positive sense—the fact that you *can't* do this in the lab, even though people who do intelligent design will say, "Ah-ha, see, irreducible complexity." It might in fact be that it was very complex in the sense of requiring a long and complicated process or computation to arrive at.

CHURCH: What we can do in the lab, though, is to reconstruct intermediates and characterize them and say, "OK, here's something we found valuable in the lab that has fewer proteins, a slightly different reaction," and make a plausible timeline to say,

"OK, given that there were 1044 water molecules on this planet and we can't reconstruct that in the lab, maybe there was a fairly small number of environments that actually did the trick," and if we construct intermediates that are convincing, then we could do small pieces of that pathway in lab timeframes.

LLOYD: Well, that would certainly be my hope. I'm just saying that that's a hope.

VENTER: But the power of selection can give you at least 7 to 8 orders of magnitude of selectivity.

LLOYD: Sure, absolutely. It's definitely worth doing. Absolutely, yes.

VENTER: That's what came out of the phi X work: you can make 106, 107 different molecules out of the assembly, and if they're viable you'll select them.

CHURCH: But Seth was saying that if we try to do the whole process, from primordial soup to ribosomes, we haven't got 108 leaders times 109 years to do it.

LLOYD: You're trying to reproduce this metabolic phase that Freeman was talking about, and in some sense once you're already at the digital phase, as you yourself said, life is not just genes, it's the machinery required to take those genes and then reproduce them, which means viruses and cells, and somehow monkeying with the program might be easier than creating viable new programs from analogs of things that are out there. This seems to me potentially easier than trying to construct this process that led up to ribosome, when you don't even know what it was in the first place.

DYSON: You have to look for something that components of ribosomes might have done in order to evolve.

LLOYD: Right. There's the example of rhodopsin, which you came up with, which provides an inspiring example. It could easily be that rhodopsin showed up as an earlier version of photosynthesis—less efficient, but hey, still good for harvesting energy—and then only later, "Oh, look, it also could be used as a sensor!" discovered by natural selection. Somehow, natural selection is full of all these tricky little switches, which makes it very hard to trace back what happened.

VENTER: Well, we can trace back to where it switched from light to chemicals and became the key driver of nervous systems.

LLOYD: Actually, Bob, I'd be interested in what you think if you had to bet on the success of this particular venture—of trying to re-create a ribosome from scratch, trying to come up with a pathway.

SHAPIRO: You can synthesize in the laboratory a ribosome from scratch, undoubtedly.

CHURCH: You mean evolve a ribosome?

LLOYD: Evolve a ribosome.

VENTER: We have synthetic ribosomes in our lab; they're just not totally efficient right now. We didn't design them; we're copying the design.

SHAPIRO: What I would say, and Freeman is probably in my camp, but I hear as I listen around the table, example after example of what I call DNA-centric thinking. Of equating life with DNA. My problem is, I know too much about DNA. I spent my life in DNA chemistry, and to me it looks like a highly evolved organism. Life started without DNA, without RNA, and undoubtedly without proteins, and was yet alive.

Life undoubtedly had to start with what nature gave us, and there's a different approach, called the bottom-up approach, where you try to use physical principles and ask what would what we regard as inanimate matter do when subjected to an appropriate environment and a liberal supply of free energy, and what combinations of those might work to kick off the living process.

Now, we're an example of one successful conclusion of the living process, but not necessarily the only example, nor need life necessarily have our exact set of components. There's a famous set of experiments from about 10 years ago, when Albert Eschenmoser, a brilliant Swiss synthetic chemist, set out to prove why nature had to select DNA. With enormous Swiss skill and manpower, he set students to making DNA-like molecules using different sugars, one after the other, expecting that in every instance he would fail. But in fact he succeeded, and he found that different sugars in many cases were superior to DNA. They had greater stability; they had fewer complications in replication. There's PAN, and TNA—there's endless ones, and so to me DNA is probably what evolution stumbled on through accident, and it's the easiest thing that could be come upon by slow trial and error and that would make a molecule that could be replicated by proteins, and that's how it came into being. Now, to me, [the thing to do,] first as an imagination experiment but ultimately in laboratory experiments, would be to try and see where else, starting from simple chemicals and energy, you might go in the direction of evolution.

DIMITAR SASSELOV: I will start the same way, by introducing my background. I am a physicist, just like Freeman and Seth, in background, but my expertise is astrophysics, and more particularly planetary astrophysics. So I'm here to tell you a little of what's new in the big picture, and also to warn you that my background basically means I'm looking for general relationships—for generalities rather than specific answers to the questions we're discussing here today.

So, for example, I am personally more interested in the question of the origins of life than the origin of life. What I mean by that is, I'm trying to understand what we could learn about pathways to life—or pathways to the complex chemistry we recognize as life—as opposed to narrowly answering the question of what is the origin of life on this planet. And that's not to say there's more value in one or the other; it's just the approach that somebody with my background would naturally take. And also the approach that's in need of more research and has some promise.

One of the reasons there are a lot of interesting new things coming from that perspective—that is, from the cosmic perspective, or planetary perspective—is because we have a lot more evidence for what's out there in the universe than we did even a few years ago. So to some extent what I want to tell you about here is some of this new evidence and why it's so exciting, in being able to inform what we are discussing here.

I want to, first of all, convince you about three things that are important to my approach. The first is that what we're looking for is baryonic in nature. What I mean by that is something of which I don't need to convince you, I believe, but you should bear it in mind, because this is a feature of our universe, the one we observe. Baryons are all the particles that make up atoms and all that's around us, including ourselves. But that's not necessarily the most common entity in the universe; you—I'm sure—know about dark matter and dark energy. I think we have to agree that what we're looking for and would call life is baryonic in nature, and there's good reason to believe that dark matter and dark energy are not capable of that level of complexity in this universe yet—or at all.

The second point I want to convince you of—or use as my background for what I'll tell you here—is that we should agree that what we are looking for, what we call "life", is a complex chemical process: basically, the ability of those atoms to combine in nontrivial ways. This is my point of departure; I look at life more from the purely thermodynamic aspect—that is, from the point of view that Robert described and Harold Morowitz has been eloquent in defining and done some research on. That is, what is the parameter space in which you can have chemistry complex enough to lead to a qualitatively new phenomenon, a phenomenon we don't see in the rest of the universe? That's an important point here.

Do we know enough about the universe that we can have sufficiently good feeling about that parameter space? Obviously we don't have detailed knowledge of most of the observable universe, but the last 50 years have seen a revolution in that field, in the sense of the ability to get diagnostics of very distant objects and a very large number of objects.

The databases in astronomy, up until just a few years ago, were larger than what biology had. It's only now that biology has exceeded that. But one aspect of these databases is that you rarely see unusual, unexplained phenomena. Despite what you all would like to write on the front page of your newspapers, there's a lot of very boring data there, which is hundreds of thousands, millions, of stars that have exactly the same isotopic and chemical patterns predicted by the theory, which is well developed and is called stellar evolution (although it has very little to do with evolution as the term is used in biology).

But it's one of those steps we now understand as the development of our world—that is, of our universe: of starting with simple baryonic structure for matter, which then becomes more and more complex. Stellar evolution is one of those phenomena that did not exist in the first half-billion years of the universe. And this is not a hypothesis; we know it. We actually can observe a lot of it, and we know there were no stars during the epoch of recombination, which is the cosmic microwave background with all the structure we see in it. And then there were stars, and then stars started a new process, which is the synthesis of the heavy elements. That is, baryons working together as elementary particles and building a structure—the Mendeleev table, which then would lead to chemistry.

13.7 billion years ago is where we see the precursor of the microwave background radiation, so that's our first well-studied piece of evidence. Then about half a billion years later is the time when the first stars form from the gas, and they're mostly made of hydrogen and helium. Then they go through a period where, over 5 billion years, they produce enough carbon, nitrogen, and oxygen and all the heavy elements—where you start effectively producing planets. And then we come to 4.5 billion years ago, to the origin of our own solar system and Earth. And almost within a half-billion years, some complex chemistry we now see covering entirely and co-opting the geophysical cycles of this planet. So that's to give you a quick idea about the time scales.

In that sense, life is an integral part of the global development we see. And although we know only one example of it, it doesn't seem unusual when you think of it that way—as a progression of complexity that the baryonic matter in this universe has the propensity to lead to. So the question then is, How is this good for understanding the origins of life or possible pathways? And even more generically, could we design experiments in which we can find out whether all these possible baryonic pathways merge into one—the one that produces life here on Earth—or are there multiple pathways? Even if you could answer that question, it would be very exciting, because it will tell us something about the general rules of complexity that baryonic chemistry can lead to.

The third aspect I want to convince you of is that we know quite a bit about the universe, and there are only a few places in the universe where that complex chemistry is able to survive over a sufficiently long period of time. And vacuum is not one of them, in the sense of surviving in which you were talking about the origin of life: starting with smaller molecules, which then have enough time to lead to more complex ones. And when I think of vacuum, I don't mean the surface of a comet but the interstellar medium, with its very low density.

I can imagine life that started on some surface then migrating to live in the interstellar medium. But I cannot imagine, as an astrophysicist, that there is an environment stable enough over the time scales necessary for that chemistry to take place. So I am a little biased, in that sense, toward planets and planetary systems as the only environment we know of today, as far as we know in the universe, which has all of those factors—that is, stability over long periods of time and sufficiently low or moderate temperatures. Stars are stable over billions of years, but they all have very high temperatures. Basically, the overall thermodynamic window that Morowitz talks about, which allows complex chemistry. That's actually a much broader requirement than simply having water.

When people talk about habitable environments, sometimes they equate that to the existence of water, or the ability of water to be in liquid form. But whatever your idea of what habitable is, the bottom line is that there are not that many objects or places in the observable universe that allow that. In fact, planetary systems are certainly not only the best but probably the only places where we're certain that complex chemistry can occur.

Then the question is, How much do we know about planetary systems? Up until 12 years ago, essentially we knew of only one: our solar system. That situation is similar to what we have with life; we have only one example. And that's bad from many points of view, and astronomers learned it the hard way, because it turned out that what we had theorized about planets was very solar-system-centric, and we missed a lot of things we should not have missed. But that always happens when you have only one example of something.

What planets allow you to do, now that we know how many different types of them there are, is provide a pretty good estimate of what to look for. One of the things we learned is that we do not necessarily have to look for planets just like Earth. In our solar system, we have a variety of planets: You have Jupiter, which is much bigger than Earth, 10 times in size, 300 times in mass; you have Saturn; you have Neptune and Uranus—all giant planets, all made of gas—then you have very small planets, Earth,

Venus, Mars, Mercury, and then comets and asteroids. There is a significant gap in masses between 1 Earth mass and 14 [Earth masses], where Uranus and Neptune are. As we would say in physics, it's more than an order of magnitude. And it allows for a whole set of phenomena that could happen in that range and which we've been missing. From what we understand now—both from theory and more recently, in the last 2 years, from observations of such systems—the fact that our solar system has no planet like this is a fluke. It just happened, the way the planets were formed, that what ended up being our solar system has no planet in that mass range. The majority of planets in that mass range will be like the Earth, and for lack of a better term we ended up calling them Super Earths.

I get a lot of flak for introducing that term, but it comes from my bias as an astronomer. We call stars that are bigger than giants "super giants"; we call stellar explosions which are more energetic than novae "supernovae"; so it just made sense that if you have a planet larger than the Earth but otherwise in essence similar to the Earth you would call it a Super Earth.

Now, why is that interesting? If you limit yourself to planets larger than Venus and Earth but not *much* larger than Earth, then you're left with very small numbers in the galaxy as a whole, and in our part of the galaxy as a whole. But if you allow yourself to count Super Earths as part of the inventory you can tap, then your numbers grow by 2 orders of magnitude.

LLOYD: What's the concentration of the smaller ones? What fraction of solar systems, or stellar systems, have "sub-Earth" planets?

SASSELOV: Ah, that's a difficult question, because they're hard to see. We have some estimates, which go to about the fraction of an Earth mass. Well, let's just say 1 Earth mass; we have no technical evidence for less than that. That's from a technique called microlensing, by the way. The evidence for this is in part statistical, but that's quite often the case—you observe many objects and you build statistical cases.

On the one hand, we already have detected a number of Super Earths—the current number is five. Despite the difficulty of detecting smaller planets, you are detecting an increasing number of those in the planetary systems you are observing. In other words, as you go to smaller and smaller masses, below about 12 to 15 Earth masses, the numbers actually rise, despite the statistical biases of having fewer of those. As our technology improves—which, by the way, it is, on a monthly basis—we will be discovering more of them.

Microlensing is sensitive to the entire mass range of planets, all the way down to 1 Earth mass, and actually in fact a bit smaller than 1 Earth mass. This technique is scanning without any prejudice a large number of stars, and to this point they have actually detected more Super Earths—smaller planets—than larger planets. Which then tells you that if you take the current statistical numbers, which we have already figured out pretty well, because we have larger planets in large numbers from the last 12 years of study, you can actually estimate what the expected number of such smaller planets is.

There is a third line of evidence, which, being a theorist myself, I would not push too hard, but theoretically if you form large planets you also form small planets, and there is no particular theoretical prejudice that you will somehow create gaps like the one we have in our solar system, where you have only very small planets and very big planets. So the final question here is, Are these Super Earths any good for what we're interested in?

VENTER: What's the number in the universe of Super Earths?

SASSELOV: Let's take our galaxy as an example, not the whole universe. We now have a pretty good idea that there are about 2 or 3 times 1011 stars in the galaxy. Of those stars, about 90 percent live long enough for the kind of complex chemistry we have in mind—half a billion years or longer. However, only about 1/10 of these stars have enough heavy elements so that planets will form around [them; Otherwise, planets] will either not form at all or will have a significant deficiency—in fact, we have evidence for that. Then the question is, How much do we know about the number of Super Earths? Basically of those left over, where we have 10 billion or so, you would say that it's only a fraction which is less than 50 percent but larger than 10 percent from those arguments that I gave to you so far.

And then you look where in the planetary system you are. You don't want to be right next to the star, and you don't want to be too far from the star—this is following Morowitz's thermodynamic estimates for the temperature range. The bottom line you end up with is about 100 million planets that I would call habitable, in the sense that they allow this kind of complex chemistry somewhere near their surface—100 million in our galaxy.

VENTER: And how many galaxies are there now?

SASSELOV: Oh, that's a large number, but it's similar to the number of stars. I insist on doing [the estimate] for the galaxy because I'm interested in the experiment; I'm a theorist, but I trust the experiment. How many of those environments can we study soon enough—while I am still alive—and with enough detail that we can help you guys, the chemists and the molecular biologists, to constrain your experiments on those pathways to life. Basically, the estimate is many. Because if you have 100 million [habitable planets] in our galaxy, then only in our vicinity, with the experiments already under way, we'll have at least 50 to 100 in the next 5 years—50 to 100 for which we can get some data that will be interesting to inform those questions.

VENTER: So your data set would exclude things like [Jupiter's moon] Europa?

SASSELOV: No, not at all. Europa is a great place to look for life. But the reason Europa is viable is because of Jupiter. If Europa were just by itself, we may not consider it that viable. I'm trying to be conservative here. But there's another reason why I talk about the 100 or so [objects we'll] be able to study: This is because I want to be able to study them *outside* our solar system. And the question is, how do you study a Europa in a planetary system 50 light-years away? Very difficult. But can you study a planet 5 times more massive than Earth and twice the size of Earth? Yes. Even much more easily than an Earth-size planet.

So the point I'm making is that the fact that Super Earths are viable as planets in comparison with Earth is great for our ability to do these experiments, because it's much easier to detect and study a planet that is twice as big as Earth and still viable. You can learn a lot from it.

One of the reasons I call these planets viable—and in fact even more viable than Earth—is because they have the basic characteristics of Earth except in a much more robust way. You probably know there's a big problem in planetary science, which is the comparison between Earth and Venus. Why does Earth have an atmosphere which is not very hot? That's sort of understood—not yet, but sort of. Why does Earth have plate tectonics, while Venus does not? That's not understood—or, we are at the verge of starting to understand that. These are questions that are much easier to answer for Super Earths.

It turns out that plate tectonics, as understood from Earth, is a process going on theoretically much more easily on a slightly bigger planet. In fact if you do the theory as best as you can today, the Earth is at the margin of what's viable in terms of plate tectonics. Probably some of you may know that plate tectonics is an important aspect of the viability of a planet, in terms of surface conditions, because it's a good thermostat; it keeps the climate more or less stable over long periods and also allows you easy access to the large reservoir of chemicals and gases in the mantle of the planet. In that sense, Super Earths are as good as Earth, and, I would argue, better. They have more stable and robust surface conditions. So they're as good as Earth, if not better, and they're easier to study. So we have a very bright future, as far as being able to at least find out what's going on.

VENTER: What role does gravity play in the Super Earths?

SASSELOV: A positive role. If you look at the general amount of out-gassing, fluxes that interchange between Earth's mantle and its atmosphere, Earth's gravity is close to marginal. We know Mars is an example of being definitely submarginal in retaining a sufficient atmosphere, and hence making this thermostat, being viable, and providing stable conditions over at least a billion years. Having more gravity is actually better.

VENTER: It increases the odds of having an atmosphere?

SASSELOV: Of keeping it. You always have an atmosphere—even Mercury has an atmosphere. There is some helium being punched out of the surface of the planet, but Mercury simply cannot retain any of it. It just goes away.

SHAPIRO: Which is the closest known Super Earth?

SASSELOV: In fact there are two of them: Gliese 581c and d; both of them are Super Earths and just 20 light-years away. Wilhelm Gliese was a German astronomer, 1915-1993. The c and d stand for "planet c" and "planet d". There is also "planet b", which is a bigger, Neptune-like planet. Thirty years ago, Gliese made a catalog of all the nearby stars. A lot of them are very faint, hence they were only identified in this catalog, and it's a common practice to call the stars by the name of the author of the catalog, with a consecutive number.

Someone asked if it's possible to launch a spore [the idea known as "panspermia"] and hit something 20 light-years away. The answer is yes—from the physics point of view it's possible. I have a colleague who says that if anything is possible given the laws of physics, it happens in the universe—but he's a physicist. Let me qualify that. Panspermia originated in the modern sense in the 20th century, back when there was a possibility that the universe might be older than 20 billion years, or maybe eternal. [Then it would be] much more likely that complex chemistry originated somewhere in the universe and spread; it's a robust system, as we know it on this planet; it can really spread, and there would be plenty of time for it to spread, even over the large distances and given all the vagaries of high-energy astrophysics. However, the universe is only 13.7 billion years old, and you have to subtract the 0.7 in order to have the first generation of stars. The first generation of stars were made of hydrogen and helium—there was no carbon, no oxygen, no metals. They were very large, and they could not have sustained protoplanetary disks, let alone allow planets to form. It took a long time before you could start forming small stars in which the protoplanetary disks would have enough solid particles to coagulate to form planets. That was thought of theoretically but with very large uncertainties; now we seem to get, somewhat unexpectedly, evidence for this. Very strong evidence, in fact.

There are a lot of searches for planets and planetary systems now targeted toward stars that are slightly older and have less heavy elements than our own sun. You see a precipitous drop in the detection of such planetary systems. In fact, after a certain factor, about 10 down, in such heavy elements, nobody has been able to detect a single planet, which is kind of strange, especially because there are people who are trying very hard. One of my colleagues has been trying now for 8 years and has come up with zero. And for [stars of] normal metallicity—that is, stars like the sun—the numbers are very high, already 250.

That means you have to come to that period in the history of the universe when stars like that form—stars like the sun or a little less rich in metals—simply because it takes that much time for the previous generations of stars to synthesize those elements through nuclear fusion. And we know that number: Another of the big successes in the last 5 years is that you can now go back all the way to that time and see—literally see, by measurements—the increase in heavy elements as you go from one generation of stars to the next and so on. Basically you are left, I would argue, with about 7 billion years. So the first complex chemistry that could have occurred on the surface of a planet would have started about 7 billion years ago. Now, in 7 billion years, it's very difficult to bring something from there to here, especially if you want to bring it here 4 billion years ago.

CHURCH: How hard would it be to hit [Earth]? At that arc angle, and given the radiation damage, and all the rest.

SASSELOV: We don't know that. It's easy to calculate the cross-section for it hitting. But understanding how it's going to make it here is very difficult.

DYSON: There's some evidence. A radio astronomer in New Zealand called Jack Baggaley is observing dust grains coming into the upper atmosphere, and he claims that a lot of them are extrasolar. They come from beyond the solar system, and furthermore they are preponderantly in a certain direction in the sky, from a star called Beta Pictoris, which is not very far away, 60 light-years or something like that. And it's quite a young star, but it has a large dust cloud around it. So it's plausible that dust is coming from this star and hitting the Earth. And if dust grains are coming, there's no reason at all why bigger objects shouldn't also be coming, and they would probably be following similar trajectories. So in principle we know that stuff is arriving from other solar systems. Whether anything is alive on Beta Pic is another question.

SASSELOV: By the way, the stuff coming from Beta Pic started coming to Earth only recently, in the last 100 million years or so, because Beta Pic didn't exist before that, and it took that long for it to come.

DYSON: Right, but it's easy to get here within the time available.

VENTER: We had these large asteroids hitting Earth after microbial life existed here. We splashed a hell of a lot of stuff into space from those hits. That's why I was trying to push you for a number in the universe. You know, 108 probability in our own system is a pretty high probability. There could have been a million origins [of life], all contributing to panspermic events, you know?

SASSELOV: Right, and then the question is, Can you cross from one *galaxy* to another? And that takes billions of years. That's the problem; we can't go from one galaxy to another.

DYSON: Isn't our galaxy big enough?

SASSELOV: That what I was saying. That's why I was making the estimate just for our galaxy.

DYSON: For now.

SASSELOV: Yes.

VENTER: At comet rates, how long would it take to go the 20 light-years? Comet speeds.

SASSELOV: About a million years. And I'm talking about the really fast comets, the fastest we've observed.

CHURCH: There would be a lot of radiation damage in a million years, I would guess.

SASSELOV: Sure. The problem is that we don't yet have evidence—I'm talking about these fast comets—we don't have the evidence for a comet coming from outside the solar system. All of them have close to parabolic orbits.

VENTER: How long would it take a 0.1 micron object to collect 3 million rads of radiation,

SASSELOV: I don't know.

VENTER: We can find organisms right here that could take 3 million rads of radiation.

LLOYD: And I bet it wouldn't be that hard to calculate what the velocity distribution of such organic objects would be after an asteroid hit, and roughly how many there would be going out.

SASSELOV: I would prefer a larger object in which your prize collection is embedded. Because then you shield it from cosmic rays; you shield it from any kind of radiation.

CHURCH: It has to be really large, you know, like meters.

SASSELOV: No, I think if you're just worried about the million years and a certain dose of rads, it shouldn't be larger than a few centimeters.

VENTER: Well, up to recently, we've been dumping all the feces from the Space Station into space—so that's kind of shield, anyway.

SHAPIRO: I have a different question. Couldn't we get some estimate of the probability of material leaving the Earth—splashing off, as they say—by examining the surface of the moon in protected areas? Say, those areas that are in permanent-shadow craters.

SASSELOV: Actually, several people suggested that as one of the interesting experiments to do when going back to the moon—to look for those. And we know which part of the moon would have most of it—it's not an even distribution over the surface, because of orbital dynamics. So that will be a very interesting thing to do.

QUESTION FROM THE PRESS: How would we be able to study the properties of these exoplanets? Is there any way to do it other than looking at transiting planets?

SASSELOV: We've thought a lot about that, because this is where we're spending our money right now. There is in fact an exoplanet task force, which is tasked to think this question over, and this is part of the direction they're writing into their report right now. The idea is that we want to have two parallel paths. One is the now old-fashioned one, just a few years old, called Terrestrial Planet Finder (TPF)—direct imaging, which is still doable but probably will take longer technologically. By direct imaging, we mean you're not imaging the surface of the planet directly but you are imaging the planet separately from the star, and you are able to get spectroscopic information that way, as well as some surface information; if the planet spins, then you see variations, which can be interpreted as surface information.

In the meantime, though, technologically it's much better to look for transiting planets and to study transiting planets. Because what transiting planets allow you to do is not only to discover the planets but once you've discovered them you can measure their mass and radius precisely. And by precisely I mean to an accuracy of 1, 2, 3 percent, which is very precise. That gives you the mean density of the planet. It turns out that this mean density can tell you whether the planet is really a small Neptune-like planet—that is, hiding as a Super Earth but really a gas-rich planet without a solid surface anywhere—or it's an Earthlike Super Earth—simply a version of Earth, just bigger. Once you've done that measurement, the next thing you can do is use measurements both during the time when the planet is in front of the star, which is called a transit, and when the planet is behind the star, which is called an eclipse.

In the first case, you measure gases in the atmosphere through transmission, which is like passing through the atmosphere of the planet. In the second case, you measure surface features. And the surface features give you a map of the surface: a color map if you do it in the infrared and an albedo map if you do it in the optical. Right now, technically, it is possible to do this for a Super Earth, say, with the existing Spitzer telescope. That's actually where we are putting our money right now, and we hope NASA will put money into the other one, the TPF.

PRESS QUESTION: What is the prospect for being able to measure the atmosphere of a Super Earth and saying, "Hey, this thing looks like it has an atmosphere out of chemical equilibrium; there's oxygen"—there's something there that makes you sit up and say this thing looks like a place that has life. What is the prospect for doing that?

SASSELOV: Five to 10 years, where 5 is more likely at this point, the way things are developing. If we're lucky, it can happen even in a year or two. But we have to be lucky. The projects that are going to discover large numbers of them [Super Earths] are coming up. CoRoT [Convection, Rotation and planetary Transits] is one of them, but NASA's Kepler is much more so, and there are a couple that have just started, or are being built, that will produce enough of those planets that you can cherry-pick and say, "Ah-ha, now I have a few that I can study in detail!" But in 10 years, we'll have a whole gallery of them, as opposed to just a milestone.

DYSON: Which molecules will you be looking for?

SASSELOV: Anything we can see. Basically the idea is to have enough signal-to-noise that we can see them all. The resolution is not an issue, because most molecules have broad spectral features, so it's a matter of signal-to-noise. And we'll try to see what we see, and personally that's one of the reasons I'm involved in origins-of-life research. Because I felt I would be embarrassed to have this gallery of spectra and maps of Super Earths without being able to answer the question, "What do you think is going on on this planet—is it chemistry, or could it be biology?"

DYSON: Can you see oxygen and nitrogen?

SASSELOV: Oh, oxygen and nitrogen are easier to see. Partly through their proxies, which are CO2 and CN. The molecules.

PRESS QUESTION: If you were looking at Earth from a Super Earth, is there anything we've done to the environment that you could detect? A large increase in carbon dioxide?

SASSELOV: Yes, people have done this research already—partly in preparation for the Terrestrial Planet Finder. The strongest indicators are the existence of ozone—free oxygen—and simultaneously amounts of methane. The imbalance is what leads you to believe that something unusual is going on and cannot be reproduced by any of the global planetary cycles we sort of understand. It's easier to complete the parameter space of global planetary cycles, like the carbon cycle, the sulfur cycle—and say we're outside any of that parameter space; that is, you cannot explain that combination of atmospheric gases with any of those cycles operating. So by exclusion you will see that there's something unusual here. But from that point of view, my estimate of habitable planets, 100 million in our galaxy, excludes the Earth. The Earth as it is now is not very habitable; it's a hostile environment for complex chemistry.

DYSON: But if you were looking at Earth this way, would there be enough CN to detect?

SASSELOV: No. The Earth is actually quite a difficult case, in that sense.

VENTER: So if it wasn't for the influence of religion, wouldn't we just logically assume that the extrapolation from life here to the statistical base—you know, that we will find it everywhere?

SASSELOV: Yes, I would say microbial life—that is, the complex chemistry of that sort—is very likely, and the more important thing is that we'll have some evidence to say something intelligent about it, rather than just saying it's very likely.

DYSON: Yes, it could get stuck in any of these phases. I think the phase where you have to invent ribosomes is probably the one you're most likely to get stuck at.

LLOYD: Though it seems to have happened relatively rapidly on Earth.

SASSELOV: My big question to all of you here is, "Can we do it by exclusion?" Can we develop, again, this parameter phase of chemistry to such a completeness where I can look at these 50 planets 8 years from now, and say, "Well, I know why all of these have what they have on their surface and atmosphere, but this one has none of that. It's out of equilibrium, and that cannot be explained simply by physics and chemistry; it must be something that's more complex and is potentially life."

VENTER: Ken Nealson, who is project scientist for the Mars Sample Return mission, had to think of some of these issues a lot, and he said the number-one thing to look for is the phosphate bond. That's the single greatest signal for biological life as we know it.

DYSON: In looking for life as we know it—or perhaps life as we *don't* know it.

VENTER: Might as well start with what we know.

CHURCH: How easy would it be to detect the phosphate bond?

SASSELOV: That would be very difficult. I was thinking the other way around: We understand physics quite a bit, chemistry I hope enough, and so if we say we understand chemistry and physics, and this is neither physics nor chemistry that we see there, we've got biology.

SHAPIRO: The trouble is, we're looking for a separate origin; this has the great philosophical impact. If we discovered that life was on Mars but just as a spillover from Earth, it would be a curiosity but it would not turn our view of the universe on its head. On the other hand, if we discovered life that's different enough that it couldn't have originated here, the spread would validate what he's been saying—that I've been saying: that life is inherent in the universe.

VENTER: Well, the two aren't incompatible. It could be identical to what we have here everywhere. The same chemistry and we find it everywhere.

CHURCH: But he's just saying it's hard to prove that.

SHAPIRO: Hard to prove—a muddled case.

CHURCH: Well, it's not a theoretical argument. You either find it or you don't.

LLOYD: Well, that's what's so upsetting about this work. Of course saying, "Oh look, there's something we don't understand; it must be life" is perhaps not the most compelling argument in the world. But if there *is* something weird going on, and it isn't explained by any of the models of life we have already, then that would be interesting.

SETH LLOYD: I'd like to step back from talking about life itself. Instead I'd like to talk about what information-processing in the universe can tell us about things like life.

There's something rather mysterious about the universe. Not just "rather mysterious," extremely mysterious. At bottom, the laws of physics are very simple. You can write them down on the back of a T-shirt: I see them written on the backs of T-shirts at MIT all the time. In addition to that, the initial state of the universe, from what we can tell from observation, was also extremely simple. It can be described by a very few bits of information.

So we have simple laws and simple initial conditions. Yet if you look around you right now, you see a huge amount of complexity. I see a bunch of human beings, each of whom is at least as complex as I am. I see trees and plants, I see cars, and as a mechanical engineer, I have to pay attention to cars. The world is extremely complex.

If you look up at the heavens, the heavens are no longer very uniform. There are clusters of galaxies, and galaxies and stars, and all sorts of different kinds of planets, and Super Earths and Sub Earths, and superhumans and subhumans, no doubt. The question is, What in the heck happened? Who ordered that? Where did this come from? Why is the universe complex?

Because normally you would think, "OK, I start off with very simple initial conditions and very simple laws, and then I should get something that's simple." In fact, mathematical definitions of complexity—like algorithmic information, say—simple laws, simple initial conditions, imply that the state is always simple. So what is it about the universe that makes it complex, that makes it spontaneously generate complexity? I'm not going to talk about supernatural explanations. What are natural explanations—scientific explanations of our universe and why it generates complexity, including complex things like life?

I claim that there is a basic feature of the universe that makes it natural for it to generate complex systems and complex behaviors. We shouldn't be surprised by this. It's intrinsic in the laws of physics. Not only that, we know what this feature is. Let me tell you what it is, and then I'll tell you what it has to do with life. Because the spontaneous generation of complexity is important for lots of things other than life. Remember, life is overrated. There's plenty of other interesting stuff going on in the universe other than life. Long after we're all dead, and maybe other biological forms—carbon-based forms—of life are dead, I hope that other interesting things will still be going on.

OK. What is this feature that's responsible for generating complexity? I would say it's the universe's intrinsic ability to register and process information at its most microscopic levels. When we build quantum computers, it's one electron/one bit, to paraphrase the Supreme Court. Because of quantum mechanics, the world is intrinsically digital. That's what the "quantum" in quantum mechanics means: It says the world comes in chunks. It's discrete. And this discreteness implies that elementary particles register bits. Their state can be described by a certain number of bits. In the case of the electron spin, one bit. In the case of photon polarization, one bit of information. Bits are intrinsic to the way the universe is. It's digital. And this digitality at the level of elementary particles gives rise to a digital nature for chemistry, because chemistry arises out of quantum mechanics together with the masses of the elementary particles and the coupling constants of nature and the electromagnetic force, et cetera.

Quantum mechanics means that there are only a discrete number of species of chemicals. You can put together 2 hydrogens and an oxygen to make a molecule in only one way I know of. This means we can catalog chemicals in a discrete list—chemical #1, chemical #2, chemical #3—you can order it any way you want according to your favorite chemicals. But it's discrete. This digital nature of the universe actually infects everything—in particular, life. It's been known since the structure of DNA was elucidated that DNA is very digital. There are 4 possible base pairs per site, 2 bits per site, 3.5 billion sites, 7 billion bits of information in the human DNA. There's a very recognizable digital code, of the kind that electrical engineers rediscovered in the 1950s, that maps the codes for sequences of DNA onto expressions of proteins. There's a digital nature to the universe, and quantum mechanics makes this happen.

But the digital nature of the universe doesn't immediately tell you why the universe is complicated and why something like life should spontaneously arise. The fact that we're here doesn't tell us anything about the probability that life exists elsewhere in the universe. Because we're here, and so we have to be here in order to contemplate this question—this tells us nothing about the probability of life except that it can exist. That's why this kind of question that Dimitar is trying to answer by looking for planets and signatures of life elsewhere is so important. We really don't know how likely it is that life should arise.

So why does complex behavior arise? Well, the universe is computing at its most microscopic scales. Two electrons, 2 bits of information, every time they collide, those bits flip. It's just these natural interactions and information processing that we use when we build quantum computers. Now, I claim—and I can claim this because this is a mathematical theorem, which is different from just mere observational evidence—that when you have something that's computing and you program it at random, just tossing in little random bits of programming, it necessarily generates complex behavior.

Einstein said, "God doesn't play dice with the universe." Well, it's not true. Einstein famously was wrong about this. It was his schoolboy howler. He believed the universe was deterministic, but in fact it's not. Quantum mechanics is inherently probabilistic; that's just the way quantum mechanics works. Quantum mechanics is constantly injecting random bits of information into the universe. Now, if you take something that can compute, and you program it at random, then what you find is that it will spontaneously start to generate all possible computable things. Why? Because you're generating all possible programs for the computer as you toss in information at random.

In fact the universe is computing. I know this because we build quantum computers—in addition, I can see a computer over there, so the universe clearly supports computation. And if you program it at random to start exploring different computations, if you go out into the infinite universe (observational evidence suggests the universe is infinite), then somewhere out there every possible computation is being played out. Every possible way of processing information is occurring somewhere out there.

OK? I don't think this is controversial, but in some funny way it seems to get people's dander up. The fact that the universe is at bottom computing, or is processing information, was actually established in the scientific sense back in the late 19th century by Maxwell, Boltzmann, and Gibbs, who showed that all atoms register bits of information. When they bounce off each other, these bits flip. That's actually where the first measures of information came up, because Maxwell, Boltzmann, and Gibbs were trying to define entropy, which is the real measure of information.

What happens when you have a computer being programmed at random? The computer generates all possible mathematical structures, and one of the most important things it does is to generate other computers amongst these structures. As first proposed by Alan Turing in the 1930s, a universal computer is a device that can simulate any other computer. It can be programmed to simulate any other computer, in a simple fashion. Including itself.

If you program a computer at random, it will start producing other computers, other ways of computing, other more complicated, composite ways of computing. And here is where life shows up. Because the universe is already computing from the very beginning when it starts—starting from the Big Bang, as soon as elementary particles show up. Then it starts exploring—I'm sorry to have to use anthropomorphic language about this. I'm not imputing any kind of actual intent to the universe as a whole, but I have to use it for this, to describe it. It starts to explore other ways of computing.

Now remember, chemicals are digital. There are only certain chemicals that can exist, and the laws of chemistry are set catalogs of chemical reactions—potentially infinite in extent because the total number of possible chemicals can be extended as much as you want. You can make polymers longer and longer and longer. You can think of the laws of chemistry, which are in some sense simple, being implied by quantum mechanics as being a catalog of this huge set of possible reactions, where if I produce chemical A and chemical B, and I put them together, that produces chemical C in abundance. Or if chemical A and chemical B are there and chemical D is also there, then chemical C is not produced.

Now, you can see the relationship of these kinds of reactions to logic, right? If A and B, then C; if A and B and D, then not C. I'm simplifying chemistry, of course, because there are temporal dynamics as well. But those dynamic if-then statements—the digital statements that lie at the bottom of computation—are an intrinsic part of chemistry.

The digital logic inherent in chemical reactions is extremely important in biology of course, because this is how the metabolism of a cell works. I receive this chemical and this other chemical; therefore I'm going to open this switch over there and turn up on this other chemical pathway. Chemistry has this computational nature embedded in it, which it inherited from the underlying computation that's going on in quantum mechanics in general. Chemistry itself, then, explores out there in the universe all possible combinations that are out there in the universe. Chemistry explores all possible computations, all possible things that could happen—including all other things that a computer can do.

Let's produce this self-reproducing structure and then see what happens. Or let's see what happens when we produce this structure and this other structure and they react with each other—let's see what they produce. We don't know exactly what went on in proto-life, but we do know the sorts of things that go on in proto-life even without knowing the exact chemical reactions that took place. It is not surprising that chemistry should produce more and more complicated structures, which then interact in more and more complicated ways, and go on to fill out more and more of the set of all possible chemical reactions, and then produce further computationally complicated structures, like, say, bacteria, or human beings, or computers.

Because there is an intrinsic capacity built into the laws of nature: this ability to process information in an open-ended fashion. And once things start doing that, then they're very hard to stop. I call such things "complexors", because they generate complexity automatically. From the mathematical or physical perspective, complexors are actually rather simple, because all they are is something that can compute which is systematically exploring a wide variety of, or all, possible computations. Once you have such a thing—once such a thing gets popped into existence, set into motion—then it will produce complexity whether you want it to or not.

We already know that at its most microscopic level the universe possesses this computational capacity because we're building quantum computers every day. In these quantum computers, we store bits of information on individual atoms, we use the laws of electrodynamics to process information in a complicated fashion, and then we get even more interesting complicated behavior, like chemistry. We shouldn't be surprised at this complexity. This ability to produce complexity infects the universe at ever higher and higher levels.

What are the implications of this intrinsic capacity of the universe to generate complexity? There are a bunch of concrete implications. Let's start by testing hypotheses for the origins of life. The first thing this capacity suggests is that since we know, to a high degree of accuracy, a large fraction of reactions for simple chemicals, we can explore the consequences of those reactions. As Bob was just telling us, we don't know a lot of these reactions when we start to include interactions of various minerals. And that's true: We don't necessarily know what all the key reactions are, and I don't think we should hope right away to be able to show how life exactly started on Earth—or elsewhere, if it started elsewhere.

But we have a good chance of showing that something *like* life should start. If we start with the set of chemical reactions that we know—and we could guess, where we don't know what they are—and we try to drive them in different ways, we would expect to see, from this computational ability, that we start out with a simple set of chemical reactions, then they start to produce more complicated chemical species, which then autocatalyze, or catalyze sets of more complicated reactions. So you'd see these species turning themselves on and then turning themselves off as they get consumed by later chemical reactions.

What you would hope to see, as this effective computation proceeds, is that it would become more and more complex as time went on; and eventually more stable sets of reactions—for instance, the citric-acid cycle Harold Morowitz is so fond of—would establish themselves as the dominant modes of operation. If we saw that happening, that would be powerful evidence for how life occurred. You would not expect to reproduce the exact origins of life, because (a) there are many possible sets of initial conditions, (b) the set of reactions could be driven in many different ways, (c) we don't know what these conditions are, (d) there's a huge number of possible ones (e) because these interactions are nonlinear and hence (f) chaotic in lots of cases, so that (g) they can be very sensitive to these initial conditions. You'd have to get very lucky to find the right ones right away. But you could establish that things like life could occur.

Just as important, you might also be able to establish no-go theorems. If we involve only a certain set of chemical reactions, it's not large enough to be computationally universal. It can extend only a certain amount and then it's just going to produce uninteresting things, such as ABABABABABABABAB—that's all it will produce. It will never produce a varied and intricate set of outcomes. And that you can analyze by looking at the set of reactions and saying, "These reactions alone are not enough." Hence, if you look at your planets and say, "Hey look, this is what's going on on this planet," then we could say, "OK, sorry—no life there."

There are lots of interesting things for life itself that we could look at. One interesting consequence—there are both good and bad consequences here—is that something like life, or the things that come afterward. . . . You're number 7, and there could easily be 7, 8, 9, 10, et cetera, keeping on going forever, if the laws of physics allow it—that's a good consequence. But it's not clear right now, given the way the universe is, that something like life could continue forever. If the dark energy persists at the same level it is right now, then in not a very long time—100 billion years or something like that is the number that springs to mind—we're all screwed and nothing can still exist, simply because all matter will have been pushed outside every other piece of matter's horizon and it can't communicate with anything else, and that's bad.

But it might also be that this dark-energy level is continually decreasing, in which case the universe could survive forever. A chapter in Freeman's *Disturbing the Universe* was influential for me in thinking about this. He pointed out that if you're willing to slow down and get very large—you have to slow down and get fat, essentially—then you can still collect free energy essentially forever and keep on metabolizing and growing. But it would require different technology than just ordinary biological life.

That's the good news. The bad news, at least from the standpoint of a scientist, is the very feature that makes complex behavior arise spontaneously. The fact that something capable of computation will spontaneously generate complex behavior means that it's also not, in general, possible to calculate (a) whether it will do so in a particular circumstance or (b) how likely it is to do so.

In fact, trying to figure out the possibilities for events early on in proto-life, just given the information we have today, is intrinsically a hard problem. If we're lucky and the pathway isn't so long, we could figure it out. But if the pathway is very long—and given the complexity of ribosomes and the way life is organized right now, it smacks of being a long and complicated and arduous process of evolution at the metabolic level, prior to the individual level; and that means it could be very hard to figure out what happened—that's potentially bad. On the other hand, there is a good thing, which is that there's a way to find out what's going to happen in life's future: that is, wait and see. I suggest we do that.

That's all I have to say. I can tell you why there's probably not life in dark energy. Or why there's not life in the first fraction of a second of the universe. But that wouldn't be very interesting.

The bizarre thing about the universe is that we understand the origins of the universe much better than we understand the origins of life. It's a simple system—we've nailed down the fact that 13.7 billion years ago first this happened and then this other thing happened, and this happened, and this happened. That's why Dimitar can speak so confidently about how stars behave; it's really, really well known. Whereas with regard to the first set of chemical reactions that started life, we just don't know what they are.

Even though every atom carries information around with it, in the Big Bang most of the computation that's going on is pretty uninteresting: It's just a bunch of stuff in thermal equilibrium bouncing off each other. To get interesting things to happen, you need the source of free energy. For that, gravitation has to kick in and take things out of thermal dynamic equilibrium.

DYSON: Yes. One of the laws of physics, which is absolutely crucial, which you didn't mention, is the fact that objects bound together by gravity have negative specific heat.

LLOYD: That's certainly important.

DYSON: That is absolutely crucial. If everything has positive specific heat, as the 19th-century scientists believed, it means that hot objects lose energy to cold objects. You're constantly losing free energy, and as hot objects lose energy they become cooler, and cold objects gaining energy become warmer. Everything goes to a uniform temperature and the universe dies and life cannot persist. That was talked about a great deal in the 19th century; they called it the "heat death," when everything goes to thermal equilibrium so life couldn't persist. But it happens that gravity has the opposite effect—that if you have an object like the sun that's held together by gravitation, that in fact the more energy you give it, the cooler it gets. And the more it loses energy, the hotter it gets.

LLOYD: Yes. If you look at star clusters, they occasionally will kick out a star, and the star will escape to infinity. And if you then look at the other stars, they're huddled together more and they're moving faster. They've gotten hotter, effectively.

DYSON: It means that in fact energy flows from cold objects to hot objects, if they're bound together by gravitation, so that you get further and further from equilibrium. That's the basic reason why the laws of physics favor heterogeneity rather than homogeneity.

LLOYD: Yes, absolutely, that's extremely important. And indeed, it's not clear how far that will go, with this historic dark energy out there. It could be that dark energy is quite useful. We just haven't figured out what to use it for yet. Of course that's the key, if you want life to survive forever; you have to do some tricky stuff to harvest energy from farther and farther away. If you take things and move them closer together, then you can take the energy out of them as you move them closer together. Of course if you do that too much, then they form black holes, and they're not as useful.

DYSON: Black holes are essential, because they are sinks of entropy; you can throw entropy into black holes and it disappears.

LLOYD: That's the cosmic garbage problem we were talking about before—the ultimate in recycling.

TING WU [Department of Genetics, Harvard Medical School] One of the things I find life so extraordinary at is self-correction—of chemical reactions as moving around certain pathways that are fairly predictable. Not that this would define life, but it's part of many lives. It will go down a pathway and it can self-correct. The most dramatic example is when DNA errors are corrected. There's a directionality there that isn't easily explained just by a chemical reaction. I don't like to anthropomorphize either, but it is as if life had a behavior—I shouldn't say "a direction," but it's moving along a direction that may not be easily explained. I was wondering if you could comment on self-correction or self-righting behavior—as a chemical reaction or not—which reminds me that as we try to define life, probably the most puzzling part of life we don't have a grasp of is behavior. So maybe we're missing one of the key aspects. I know as a biologist that behavior is almost a complete mystery right now.

LLOYD: Interestingly, this DNA-correction mechanism you allude to lies at the very beginning of my own field of quantum computing. In the 1970s, Charlie Bennett looked at the thermodynamics of this DNA-correcting mechanism, and when you're correcting errors you have to throw information away, because afterward you want the DNA to be in the right state, independent of whatever error happened before.

WU: Whatever "right" is.

LLOYD: Whatever "right" is. In this case, the DNA-correction mechanism is detecting to see "OK, do these two strands match," for instance, "or are they complementary to each other?" And if they're not, then you go back and you try to rewrite them. Then the information about the error goes away, and it turns out that this has to generate entropy, because the laws of physics at bottom are reversible. They're only irreversible in the macroscopic sense—and that means you can never throw information away for good. So if I throw information about the DNA away, that information has got to go somewhere else. And so these interactions are entropy-generating: You have to supply them with a source of free energy and drive them along. In fact, if you supply them with too little free energy they'll go back in the other direction and they'll generate errors. So an error-correcting mechanism, if it runs in the wrong direction, is an error-generating mechanism, which is actually also—not to anthropomorphize it—kind of human behavior. The ability to operate in a stable robust fashion in the presence of noise and errors is a key aspect of life and is not so easy to effect. Particularly at the level of individual quantum, may I say.

Let's now look at this question of behavior. This computational issue—the fact that things are computing, are processing information, according to things like "if-then" statements—can be thought of as the origins of inscrutability of behavior, either of chemical reactions or of human beings. Let me phrase this in terms of computers, because then I'm on safe ground because this is a theory I can prove. There's a famous theorem in computer science called the Halting Problem, which Alan Turing first proposed. He pointed out that just from the very fact a universal computer can simulate itself, that you can construct self-contradictory statements. As a result, certain questions can't be answered by a computer. One such question is, "If I change this one bit in this computer program, then will it stop and give an output?" This problem is called the Halting Problem because there's no way to compute what will happen when you set a computation in motion, other than actually waiting and seeing what happens. "There aren't any shortcuts" is another way of saying it. If something's going through a complex computation, there's no logical shortcut that lets you figure out what it's going to do, other than going through the computation and seeing what will happen.

What this means is that computers are intrinsically inscrutable. When you press "Return" today, everything should look exactly the same as when you did it yesterday. Today you press "Return" and your computer crashes. Right? Yesterday it printed out your manuscript, today it crashes and takes your manuscript with it. Has that ever happened to anybody here? It has certainly happened to me. This is a necessary part of digital computation. There's no way, in general, if a computer is performing complicated computations—and those computers are performing pretty complicated computations—to figure out what will happen except to do it. This also holds for chemical reactions, because these chemical reactions have the same sort of "if-then" quality that computations have. That's a simplified version of chemical reactions, of course, but the more complex version is at least that complex. It's at least as inscrutable. Even in the simplified "if-then" picture of chemical reactions, the outcome of a complex set of chemical reactions is by necessity inscrutable. The only way to figure out what's going to happen, in general, is to let it go and see.

This is why, if we're going to figure out what the origins of life are, we're going to need either to do some pretty major experiments and/or burn a whole bunch of super-computer power, because the only way to figure out what they'll do is to see what happens. And if it's true of computers and chemical reactions, it's certainly true of human beings. If I think of what makes other people inscrutable—I'll just speak for myself; maybe you find me completely transparent—I find most interactions with other people inscrutable. Or even interactions with myself.

If I want to see what I'll do tomorrow, I'm a free agent, and I'm the only one who will determine what that is. But the only way for me to figure it out is to go through the thinking process and then to figure it out. The inscrutability of my own actions comes in part from this essential logical feature—that the only way to figure out what will happen in a computing system is to go through the computation. And certainly for other human beings who are at least as complicated as I am, I cannot model what's going on inside their heads, and even if I could, the only way to figure out what they were going to say or do would be to go through the complete thought process they're going through. Which I just can't do.

I would say that computers and chemical reactions share with human beings the feature of inscrutability of their behavior, and there's nothing to be done about it. There are things you can try: You can get more familiar with them, you can try to model them better, but you'll never eliminate the uncertainty and essential inscrutability, because it just is the nature of anything that's behaving in a logical fashion. Bizarrely enough, it's like Spock: the Vulcan code makes him more strange and hard to understand than if he were actually irrational. It's rationality that makes us inscrutable, rather than irrationality.

PRESS QUESTION: How do you avoid the Gödel trap, in the sense that there are things that exist that you can't possibly explain the origin of?

LLOYD: Exactly. The Halting Problem and Gödel's theorem are essentially the same problem—they're very closely related, and Turing knew about Gödel's work when he came up with the Halting Problem. In fact, he came up with the Halting Problem and the Turing machine because he wanted to write about Gödel's work.

Gödel's theorem is basically the Cretan-liar paradox, which comes from St. Paul's letter to Titus, who's going out to preach to the Cretans, and St. Paul says, watch out for those Cretans—one of their own philosophers says all Cretans are big bellies, gluttons, and lascivious liars. The question is, How do you treat someone who says, "I am lying no matter what I say." And in the logical sense, this becomes a statement. Probably the best one—which is what Gödel used—is to construct a statement that effectively says, "This statement cannot be proved to be true." So it's a logical statement within a set of axioms. And there are two possibilities: Either the statement is true or it's false.

Let's say it's false; if it's false, then the statement *can* be proved to be true—but now you've proved a false statement to be true, and that's really bad, because if one false statement can be true, then you can prove all false statements to be true. As my children demonstrate to me all the time, "Dad, you just said . . ."—therefore you're unreliable in all ways. The only alternative is that the statement is true but it cannot be proved.

Such a statement is one that is, as it were, inscrutable to the logical structure of the theory. It cannot be proved from the theory, but the only choice you have is to adjoin it to the set of axioms of the theory as an addition axiom. And once you've done that, then there are more statements like this—Gödel's incompleteness theorem says that no self-consistent logical theory of beyond a certain complexity—basically, complexity that allows it to compute—is complete. The theory can always be extended in a whole variety of different ways.

PRESS: It means that there have to be things in this universe which are not the result of the series of computations. In other words, they're true, because the truth in this example is something that's produced by these calculations, but you can't find the origins, you can't trace them.

LLOYD: I agree. In fact they can't be derived from those laws, because quantum mechanics says that the universe is not really a universe but a multiverse: There are different branches to the universe, in which different possibilities are explored. I would say even in some branches you could say the false possibilities are explored, where the universe is inconsistent and then ceases to exist.

PRESS: It's possible then that life could be one of those things you cannot trace the origins of.

LLOYD: It's conceivable. There has to be a kind of infinity built into the problem. Life presumably originated in some finite context, so it could conceivably be discovered. But the kinds of finite problems that are analogous to these Gödelian problems are things like NP-incomplete problems where there's a huge number of different possibilities and you'd have to explore each one to find the answer.

SHAPIRO: I just want to emphasize, lest it slip away, one point which was in the middle of the conversation, which is basically that we may never be able to capture the actual circumstances that led to the beginning of our life here on Earth, because environments may have been destroyed or circumstances changed of which there's no record, but there's every opportunity by experiment for searching elsewhere, to find what are the general principles involved in generating life.

LLOYD: Right. Suppose we start to do these experiments, both real and computational experiments, to say, OK, here's chemistry, it's doing these funny autocatalytic interactions that are a computation, in the strict definition of a computation, and we're going to explore what happens. Then suppose that as we start doing that, we find things that give rise to complicated behavior. That certainly fits your definitions, Freeman, of proto-life—steps #1 and #2—and maybe even things that are like step #3, but what we get are totally different from ribosomes in step #3. If this happens, then that, I would say, is strong evidence that we should expect to have life in all sorts of places, involving all kinds of different ways of living other than having ribosomes.

SASSELOV: That's the question of multiple vs. simple pathways to life. Just answering that question would be essential.

LLOYD: And that's quite possible—even if it's too hard to figure out exactly how life originated on Earth. This is a much easier question, I think, to answer than the question of how did life exactly originate on Earth. Because there you have to figure out the exact initial conditions for this complicated set of chemical reactions, and that's going to be hard.

SHAPIRO: And the other point of view has been very much pushed over the ages; I think George Wald once said that if you study your biochemistry text on Earth you can pass examinations on Arcturus. Which is a star somewhere out there—and this is essentially saying the opposite.

9
The Gene-Centric View: A Conversation

Richard Dawkins & J. Craig Venter

[January 23, 2008]

Introduction by John Brockman

It's not every day you have Richard Dawkins and Craig Venter on a stage together. Richard Dawkins is responsible for possibly the most important science book of the last century, *The Selfish Gene*, published in 1976, which set forth an agenda of the gene-centric, or gene's-eye-view of life, which has become the basic science agenda for biologists for the last quarter century. And without that worldview, you wouldn't have Craig Venter changing the world the way he is today. Venter led the private group that decoded the human genome in 2001. He's working on the forefront of artificial life, synthetic biology. He's traveling around the world on a sailboat, finding millions of new genes in the oceans. Most recently, his lab was responsible for transplanting the information from one genome into another. In other words, your dog becomes your cat. What we'll present first is a conversation between Craig and Richard, and then they will entertain questions.

RICHARD DAWKINS: I thought I'd begin by reading a quotation from a famous philosopher and historian of science from the 1930s, Charles Singer, to give an idea of exactly how much things have changed. And Craig Venter is a leader—perhaps *the* leader—in making that change today. Here is Singer, in 1931: "Despite interpretations to the contrary, the theory of the gene is not a 'mechanist' theory. The gene is no more comprehensible as a chemical or physical entity than is the cell or, for that matter, the organism itself. . . . If I ask for a living chromosome, that is, for the only effective kind of chromosome, no one can give it to me except in its living surroundings any more than he can give me a living arm or leg. The doctrine of the relativity of functions is as true for the gene as it is for any of the organs of the body. They exist and function only in relation to other organs. Thus the last of the biological theories leaves us where the first started, in the presence of a power called life, or psyche, which is not only of its own kind but unique in each and all of its exhibitions." You couldn't ask for a more comprehensive destruction of a conventional view than that. That is not just wrong, it is catastrophically, utterly, stupefyingly wrong. It's wrong in an interesting way, and Craig is the best person to tell us what's wrong with all that.

J. CRAIG VENTER: I feel like this is a quiz, Richard [laughter]. Richard's book *The Selfish Gene* influenced most thinking in modern biology. I actually didn't like his book, initially—I've never told him that. But I've come to appreciate it immensely. I was looking at the world from a genome-centric view—the collection of genes put together to lead to any one species—but as we traveled around the world trying to look at the diversity of biology, we came up with larger and larger collections of genes. We now have a database of roughly 10 million; that number will probably double again, this year, to 20 million.

To put it in context, we humans have only around 22,000 genes. We represent sort of a minority in the usage of genes on this planet. But I've switched, and I've come to view the world from a gene-centric point of view—in part because we're now going into the design phase. I'm looking at genes as the design components of the future, not just as interesting elements of biology. I now look at genomes as interesting composites of genes. But we have almost an infinite variety that we could put together to create biological machines of the future. Unlike that quote from Singer, chromosomes *can* exist independently. Genes *can* exist independently. They can move around independently.

Last year, we isolated the chromosome from one bacterial species and transplanted it into another. The chromosome in the species we transplanted into was destroyed, and all the characteristics of one species went away and got transformed into what was dictated by the new chromosome. It's sort of the ultimate test in proving that this is the information of biology and dictates what a cell can do and maybe even should do. This was a precursor to being able to design life—build synthetic molecules by looking at individual genes. We now have some gene families where we have 30,000, 40,000, 50,000 members—natural variants that occur in the population. And we have major problems we're trying to overcome in modern society by looking for solutions.

The first uses we're looking at is trying to come up with alternate ways of making fuel. Instead of taking carbon out of the ground, given this diversity of biology, we have thousands, perhaps tens of thousands, of organisms that can take the energy from sunlight and carbon dioxide from the environment, fix the carbon from the carbon dioxide, and also make a potential fuel—natural gas, such as methane. When we look at cells as machines, it makes it straightforward in the future to design them for unique utility. All these advances speak against that one quotation.

DAWKINS: It's more than just saying you can pick up a chromosome and put it in somewhere else. It is pure information.

You could put it into a printed book. You could send it over the Internet. You could store it on a magnetic disk for 1,000 years, and then, in 1,000 years' time, with the technology we'll have then, it will be possible to reconstruct whatever living organism is here now. This is something utterly undreamed of before the molecular information revolution.

What has happened is that genetics has become a branch of information technology. It is pure information. It's digital information. It's precisely the kind of information that can be translated digit for digit, byte for byte, into any other kind of information and then translated back again. This is a major revolution. I suppose it's probably *the* major revolution in the whole history of our understanding of ourselves. It's something that would have boggled the mind of Darwin, and Darwin would have loved it, I'm sure.

VENTER: Well, to speak to this: For the past 15 years, we've been digitizing biology. When we decoded the genome, including sequencing the human genome, that was going from the analog world of biology into the digital world of the computer. Now, for the first time, we can go in the other direction. With synthetic genomics and synthetic biology, we are starting with that purely digital world. We take the sequence out of the computer and from 4 raw chemicals that come in bottles we can reconstruct a chromosome in the laboratory, based on either design: copying what was in the digital world or coming up with new digital versions. In fact, somewhat jokingly, I can argue that this is the only nanotechnology that actually works. Biology is the ultimate nanotechnology, and it can now be digitally designed and reconstructed.

DAWKINS: There are people who are uneasy about this kind of science. They sometimes call it scientism. And there's a certain suspicion of arrogance. The phrase "playing God" has been brought up. I don't think I have a problem with that, but I think it's something we ought to take seriously. What I do have a problem with is the possible unforeseen practical consequence of some of the sorts of things that not just you but many people are doing. I suspect that the phrase "playing God" is actually a kind of— It's a bit like the boy who cried "Wolf", because accusing a scientist of playing God is obviously stupid. But what is not obviously stupid is accusing a scientist of endangering the future of the planet by doing something that could be irreversible. We may become so used to fending off idiotic accusations of playing God that we might overlook the real dangers. Do you think that's a possible danger?

VENTER: It's a real-life danger that we're facing now. I've argued that we are now 100-percent dependent on science for the survival of our species. In part, the science of today has to overcome the scientific breakthroughs of previous years. Because we have advanced internal combustion engines; because we're so good at burning carbon that we take out of the ground; we did it blindly, without any [thought of the] consequences, that it might totally affect the future of the planet.

We can replace the carbon we're taking out of the ground by renewable sources, and the best renewable source we have is energy from the sun. Over 100 million terawatts a day hit the Earth. We have cells that capture carbon back from the environment. And it turns out, chemically and biologically in the lab, we can make anything in the lab that comes out of the ground, in terms of carbon. We can make octane. We can make diesel fuel. We can make jet fuel. We can make butanol. Ethanol—humanity's been doing that forever, through simple fermentation.

These ideas are slow to catch on. People are much more concerned that there might be new consequences of engineering biology than about this potential disastrous route we're on—totally changing our atmosphere, maybe making it impossible ultimately for our species to survive. That's a far more dangerous experiment.

DAWKINS: Did I understand you to be saying that whereas the energy we get out of the ground—oil and coal—took millions of years of all those terawatts of sunlight hitting leaves in the Carboniferous and being stored, do I understand you to be saying that now, with the biotechnology you're doing, it should be possible to capture those terawatts of energy on the fly, as it were, and use them in the present, rather than [using what's been] stored over millions of years from the past and dug out of the ground?

VENTER: Exactly. What we're doing with burning oil and coal is we're taking millions of years of compressed biology, we're burning that over the course of years and putting it into the atmosphere. We can do just the opposite. We can even capture back some of that CO2. It only takes about 1 percent of the sunlight that hits the Earth daily to replace all the fuel we use, all the energy we use for transportation. These are not huge leaps. There's just been no motivation for it, because oil was cheap. We've gone through this cycle, twice now, where people rapidly pursued alternate energy sources and then the cost of oil dropped. In fact, that's my biggest concern. The price of oil is in the hands of a very few people. And if there's truly alternatives that come on the market, the cost of oil could be artificially dropped to really low prices, killing off these essential new industries. The way forward, in a political sense, is there has to be a carbon tax on nonrenewable carbon to disincentivize people from burning it. Even the Bush administration recognizes that we're in the realm of climate change due to carbon going into the atmosphere. If *they* understand it, the rest of the world can.

BROCKMAN: At the "Life, What A Concept!" meeting, Freeman Dyson basically challenged Richard by saying that evolution is now back to the prebiotic stage of communal, horizontal gene transfer, and with the interlude of what he would call the "Darwinian moment." Richard rebutted that in an e-mail, which is rather exciting reading. But the question I have is: Dyson maintains that evolution is now man-made—cultural rather than Darwinian. Is it?

VENTER: All evolution is based on selection. We, as a species, have been affecting the direction of evolution for some time, whether we wanted to or not, by changing the environment. Now we can do it in a deliberate, hopefully thoughtful fashion, by deliberate design. But that deliberate design still has to be followed by selection.

When we look at that experiment we did, transplanting a genome from one species to another, many people who try to argue against evolution on religious grounds stick to this point-mutation and selection mode, the most limited version of Darwinian evolution, to argue how complexity couldn't occur from that. But what we see with chromosome transplantation is, we can get a million changes in a species in an instant. And not only does this happen just by our work in the lab. Looking back in history, we see major species evolution from species taking on new chromosomes. When they take on a new chromosome, it's like adding a new DVD full of software to your computer. It instantly changes the capabilities and the robustness of what you can do. Our cells can do that. We have real-time Darwinian evolution taking place in our lungs. Everybody in this room has different species of bacteria in their lungs, because as your immune system attacks these organisms, there's built-in mechanisms to their genetic code constantly making minor variations, making different proteins to fool our immune system.

This is case selection by our antibodies, and our physiology is changing those. We're changing selection of the species, perhaps ones that will survive in a higher CO2 environment. As I sailed around the world, one of the most disturbing things was that we could barely go a mile in the ocean without seeing plastic trash. We did not anywhere, in a complete circumnavigation, see a pristine beach, a beach that wasn't covered with trash. But talk about a new environment: After the major tsunami, as we sailed across the Indian Ocean, all the beaches were covered with flip-flops, which turned into rafts for crabs. So we have a new habitat for crabs, as they float around the ocean on people's flip-flops. We are very much affecting evolution on our planet. My contention is, we need to start doing it in a deliberate fashion.

DAWKINS: I want to come back to John's point about Freeman Dyson. I didn't actually disagree with him all that much. The only thing I disagreed about with him was that he was talking about natural selection as though it were selection between species, which it is not. However, the extremely interesting point he made is the transition from a very early stage of evolution which was much more open-source, with bacteria copying and pasting information in a kind of promiscuous fashion, which is exactly what *we* are now in a position to do, using both genetic information, through people like Craig, and also other kinds of information—cultural information.

So there is an interesting sense in which there still is a middle phase of what Dyson called the Darwinian phase, by which what he really meant was the highly ritualized phase of sexual exchange of information, which, as I say, is ritualized as opposed to the open-source system, which bacteria still do and which human biotechnology now does. By "ritualized", what I mean is that in every generation, exactly 50 percent of the genes of a male and 50 percent of the genes of a female are put together to make a new individual. Now, that is a highly stylized, ritualized, courtly kind of genetic-information exchange, which took over from the bacterial system and caused the invention of what we call a species, because a species just is a collection of individuals who are taking part in this stately gavotte of genetic exchange. Having usurped the earlier stage of promiscuity—shoving genetic information around all over the place—we're moving back into a new promiscuous phase. However, I wouldn't write off what Dyson called the Darwinian phase. It's been going for a couple of 1,000 million years, and it's going to go on all around us, never mind what humans are doing.

VENTER: You did use the phrase "schoolboy howler."

DAWKINS: Schoolboy howler. I did use the phrase, and that was about that one point—about suggesting that natural selection is about one species displacing another species, and that *is* a schoolboy howler. A lot of people think Darwinian selection means that one species goes extinct and another species takes over. That is not Darwinian selection; that is species extinction. It's a totally different kind of process.

BROCKMAN: Speaking for Freeman, he still maintains he's correct. One interesting aspect is that in science, debate is the way people work together, the way they advance their ideas. It's usually civil. In this case, it was very good-natured. The two major German newspapers, *Suddeutsche Zeitung* and *Frankfurter Allgemeine Zeitung*, were present, and they both ran feuilleton features on the event. And one of them said if this discussion had been in Germany, there'd be riots and fistfights. But the audiences here all seem so calm.

VENTER: Let me pick up on a point that Richard was making about the simplistic notions about Darwin and evolution. In fact, what we found in the environment was one of the biggest surprises for the scientific community. Most people expected just one dominant species. What we found were thousands, tens of thousands, of closely related organisms, all basically the same linear set of genes—tremendous variation in those genes—but there was not one dominant organism. There was this community of related organisms where perhaps none of them had gone extinct—or, if they had, there were literally thousands of ones to replace them.

The problem we've had, I think, with evolution—it has been overly simplified, because we've always been looking at the visible world, not the majority of life on this planet, which is the invisible world. In 1 milliliter of seawater, there are a million bacteria and 10 million viruses. In the air in this room—we've been doing the air genome project—all of you just during the course of this hour will be breathing in at least 10,000 different bacteria and maybe 100,000 viruses. I would look closely at the person sitting next to you to see what they're exhaling.

This is the world of biology we live in, that we don't see, where evolution takes place on a minute-to-minute basis, not on the speciations of giraffes versus elephants versus kangaroos but the tens of millions of species that constantly affect the metabolism of our planet. The air we breathe comes from these organisms. The future of the planet rests in these organisms. And the question is: If we take over the design of these organisms, does that shift the balance in any way? Or is it such a small portion of what's out there that we'll only affect industrial processes, not the living planet?

DAWKINS: My vision of life is, in a sense, even more radical than that, because I would like to regard the genomes of the giraffes and kangaroos and humans that you refer to as just another set of viruses in close-knit societies. So the gene pool, I should say, of giraffes, or the gene pool of humans, or the gene pool of kangaroos is a huge society of viruses. I'm using the word loosely. I'm using the word "viruses" because the viruses you're talking about, the bacteria you're talking about, are kind of free spirits who are out there in the sea and out there in the air. But there's another whole class of them who have come together in gigantic clubs, gigantic societies, which is you and me. And so as far as a piece of DNA is concerned, there are various ways of making a living. And some of the ways of making a living are floating around free in the air and floating around free in the water. Other ways of making a living are to club together with other bits of DNA, making a genome, and influencing the phenotype, influencing the body in which they sit, to pass them on to future generations. These are just different ways of making a living. The whole of the biosphere is a gigantic collection of criss-crossing interacting DNA, some of which jumps from kangaroo to kangaroo, or from giraffe to giraffe, but via the normal route of reproduction, sexual reproduction; others of which jump around through the air or through the water. But it's all the same kind of stuff.

VENTER: In fact, the jumping, I think, is a lot farther; they can jump from planet to planet. We have organisms that can withstand 3 million rads of radiation. They can be totally desiccated. It's been shown that they can survive easily in outer space. We exchange roughly 200 kilograms of material between Earth and Mars each year. Undoubtedly, we're exchanging these organisms. It's a question of how far they can transfer. We're starting to look at the gels of space dust, to see if we can find DNA in them. These organisms, if they were shielded within a comet, within any other material, could literally last tens of millions of years, find a new source of water, and start replicating again. Our viruses can affect the universe, just not the girl next door.

DAWKINS: There's a precious beauty in the experiments you've been describing, because Charles Darwin himself did the same thing, but with transmission of organisms from continent to continent. Darwin was concerned for theoretical reasons to argue that it's possible for living things to survive long journeys in seawater or other transmission conditions. Darwin did experiments analogous to yours, in which he took seeds and showed that they could survive for long periods of time, long enough to drift across from one continent to another. It's a beautiful analogy

VENTER: I'm certain we will find bacterial life on Mars; whether it's actively replicating or not still is a question. But it won't differ from what we have on this planet.

DAWKINS: But it'll be Earth—

VENTER: Because it will either have originated there and come here, or originated here and contaminated there.

BROCKMAN: Have you thought about exoplanets?

VENTER: Dimitar Sasselov says there are 100,000 planets, just within our own galaxy, that could support life. We will find life as a universal concept. Anywhere we will find intelligent life, we will find it's a design concept. It's an electronic concept. It's an information concept. We can transfer life across the universe as digital information. Somebody else could, in their laboratories, build that genetic code and replicate it. Perhaps publishing my genome on the Internet had more implications than I thought.

BROCKMAN: When you talk about design, you're inferring that life is a technology. Would that be true?

VENTER: Life is machinery. Life is a form of technology as we learn how to engineer it and reproduce it.

BROCKMAN: One of Richard's colleagues, J. Z. Young, at Oxford, in his 1951 Reith Lecture, said we create tools and we mold ourselves through our use of them. So if life has moved from reality to a tool to a technology, how is that going to change our view of who and what we are?

VENTER: It's a question that came up at the beginning of looking at the genetic code. Many argued that we would diminish humanity by looking at our own genetic code and understanding it. That's a simplistic view. Looking at our genetic code and trying to understand how we go from the same 22,000 genes in every one of our 100 trillion cells to a John Brockman and a Richard Dawkins is far more fascinating than anybody can conjure up, I think, from any religious or poetic form. I don't think it diminishes humanity to understand it.

BROCKMAN: It sounds fascinating now. Twenty-five years from now, it'll sound, to the next generation, trite and taken for granted. Things are going to change. With this scheme of things, I don't see any place for religions. I think we're going to relate to each other differently. The whole cybernetic idea is a huge epistemological breakdown of our traditional ways of looking at each other. We go down an empirical road, until it hits a wall, and you have to rethink everything. And that's where we are right now.

VENTER: Well, it certainly changes the definition of an Internet virus. If we can have an actual virus, digitize its code, we can transmit it around the Internet and somebody else could build that same one—or, more important, a cell to make octane from carbon dioxide based on sunlight. We need to get these transfers quickly. We are a species for whom everything out of sight is out of mind. While we worry about GMOs, primarily in Europe, I worry most about the several trillion organisms that get transferred as ballast water that ships pick up in any port after they dump their cargo, and take to another part of the world, and contaminate that part of the world with all those microorganisms and viruses. This has been going on ever since ships have taken on ballast water. We are doing a cross-contamination. The experiment Darwin did, every time a ship takes on ballast water, it moves someplace else and dumps that water, moving billions to trillions of organisms and viruses around to create environments that wouldn't normally exist.

BROCKMAN: The audience might be interested in your adventures with national governments in your surveying their waters in the South Pacific.

VENTER: What John's referring to—it's almost impossible, as a modern scientist today, to do what Darwin was able to do. On his voyage on a survey ship going around South America, he took biological samples, characterized them everywhere he went. We now have international treaties stipulating that every country owns every species within 200 miles of its borders. We found, as we sailed across the Pacific Ocean with a 1-knot current that carried a million organisms per milliliter of ocean water across a border, they went from organisms that were international to becoming of French genetic heritage. And it changes the ownership. It changes the view of science—to the point where most states now don't want discoveries made and that information published on the Internet or in scientific journals. So we've gone to the extreme opposite of open-source, to where it's hard to find a country that doesn't want to block the publication of biological information that's either originated in that country or drifted across its borders.

BROCKMAN: At the "Life: What a Concept" conference, you said something about artificial life—that it's not "if" but "when" it's happening, and it's going to happen sooner than we think. So, what is the prognosis on that?

VENTER: Well, just for the record, we have not yet created a cell driven by a man-made chromosome. Based on the chromosome-transplant experiment, though, we know that that is definitely possible. There's a lot of barriers to it. There's different mechanisms in cells where—because these are in fact key mechanisms of evolution—if you're a cell swimming in the ocean and you take up not a gene but a whole chromosome from another species, and it instantly transforms what you do as a species, some species wanted to develop mechanisms to protect themselves against that. There are a lot of barriers we have to overcome. I'm hopeful that will happen this year.

DAWKINS: Can I talk a bit about some of the risks? Craig, you were just talking about the sort of almost criminal contamination of oceans when tankers release ballasts of seawater and thereby contaminate one ocean with the organisms of another. And we're all now quite used to the idea of contamination of organisms. When you go to New Zealand, you hear thrushes and blackbirds, because the early settlers felt nostalgic for British birds and wanted to bring in British birds. I mean, it's criminal. The Duke of Bedford imported American grey squirrels into Britain, and now the red squirrel is all but extinct. We're entirely used to this idea of contamination. However, what's the equivalent that we might be doing now? What if scientists of the future are unable any longer to do serious molecular taxonomy work because the scientists of the 21st and 22nd centuries contaminated genomes by introducing genes from other radically different parts of the living kingdoms?

It's probably all right as long as very, very careful records are kept. However, you could imagine a situation in the future where the rather strict separation—at least in Freeman Dyson's middle stage of evolution, the sexual phase—where, on the whole, evolution is all divergent, there's virtually no cross-contamination of genes, if humans suddenly start cross-contaminating genes so you have kangaroo genes in giraffes or melon genes in aardvarks, how are we going to do our molecular taxonomy? Won't it be a bit rather like people trying to study the faunas and ecology of New Zealand?

VENTER: Richard, that's the most naïve question you've ever asked. And I assume you're asking it to be provocative, because in fact that's the opposite of what we see happens with evolution. Viruses move genes around from totally disparate species in a common fashion. We have genes in our genome that resemble some from distant viruses. In fact, a third of our genome is basically viral contaminate. When we sequenced the smallpox genome, the smallpox genome had half a dozen clearly human-derived genes. We see bacterial genes moving in a lateral fashion from *Archaea* to bacteria to plants to single-cell eukaryotes. We do have constant information exchange across the diversity of species on this planet. I've never heard the term until this meeting, that of the "schoolboy howler," but I would put this in that category, the simplistic view of biology.

DAWKINS: Are you saying, then, that a molecular taxonomist who's trying to work out, say, the taxonomy of marsupial mammals or placental mammals would be thrown because a bacterium or a virus had at some point carried a kangaroo gene into a jackal genome or something? You're not saying that. Are you?

VENTER: We're saying that we see evidence of every branch of life in almost every genome. It depends on which gene you choose, and that's been the problem with molecular taxonomy. If you choose one gene out of 2,000 or 3,000 in a genome and try and classify on that, you come up with one answer. If you pick another gene, you get a different tree. If you look at the genome as a whole, you get a totally different answer. So, yes, we see genes moving around.

You know, the visible world and these few visible species to me are somewhat bizarre extremes of evolution. They're not the standard. But if you look at those, in the marsupial versus a platypus genome you would definitely find a clear-cut similarity. If we sequenced another mammalian genome, we would not discover a single new gene. We would discover unique *combinations* that made that mammal versus us. But we have saturated the gene set for mammals. So we can say here that the gene set of mammals, over half of those are shared broadly with other species. You can't draw a bright line at every gene, and say, "These are plants and these are mammals. These are humans and these are marsupials," because we've used—it gets back to the gene-centric view—we've used those in the random design of biology, as we will use them in the very specific design we do in the laboratory. And taxonomy is something where people sort of fool themselves by justifying what they see with their visual acuity.

DAWKINS: The overlap of mammal genes that you're talking about could come about through common ancestry. So the platypus and kangaroo genomes contain shared genes because they go back to a common ancestor. That is the normal assumption made by molecular taxonomists.

VENTER: Yes, but once you have lateral transfer, whether it's due to viruses or anything else, the tree concept of life goes away.

DAWKINS: That's what I'm asking you. To what extent does molecular taxonomy now have to be, not overthrown, but at least thought of with great suspicion because you cannot tell which genes are in common because they're from a common ancestor or because they're cross-contaminated by viral or bacterial transfer?

VENTER: We can use the genetic code to watermark chromosomes. You can use it in a secret code. Basically, what we're using is the 3-letter triplet code that codes for amino acids. There are 20 amino acids, and they use single letters to denote those. Using the triplet code, we can write words, sentences. We can say, "This genome was made by Richard Dawkins on (this date) in 2008." A key hallmark of man-made species, manmade chromosomes, is that they will be very much denoted that way.

You could, obviously, copy something that was out there and make minor variations and nobody would necessarily know. The other key tenet of what we're doing is, the organisms we're designing are designed *not* to survive outside of the lab. I don't know anybody who's advocating making a new species and throwing it into the ocean.

BROCKMAN: I'm sure some of you have questions.

QUESTIONER: I have a question for Richard Dawkins. You've known Craig Venter for quite a while, and you remember 10 years ago, when the world was too slow for him in sequencing the genome, and he said, "I want to do it myself. I want to do it quicker." Now he announces that he wants to create artificial life to resolve the energy crisis and bring the oil price down and create new forms of energy. When will he come up with the first form of energy?

DAWKINS: You're asking me a question about Craig.

QUESTIONER: Yes. I asked him already a couple of weeks ago, and he didn't say anything about the timeline. [Laughter]

VENTER: So, supposedly I told you [to Dawkins] in secret what the real answer was and you're going to reveal it now.

DAWKINS: I'm not going to reveal anything. I want an answer from Craig about the kangaroos and the— [Laughter] I think you're confusing two quite different things. Of course you can make viruses and bacteria transfer things, and we know there are a few genes that have cross-contaminated from radically different parts of the animal and plant kingdoms, but I didn't know, until you told me today—and I'm skeptical about it—that molecular taxonomy of, for example, mammals, is endangered by cross-contamination of genomes. I don't believe molecular taxonomists—yet, at least—say, "Oh, well, we can't use this gene to get our kangaroo taxonomy right, because it's clearly been imported from a rhinoceros."

VENTER: When we look at bacterial evolution, a typical bacteria will have 2,000 genes in it. Each one of those 2,000 genes has its own separate evolutionary tree, which you can construct, and none of them have the same timeline that you could put together.

DAWKINS: But that's bacteria.

VENTER: But that's bacteria. So, viruses pick up bacterial genes all the time. They pick up mammalian genes all the time. A third of your genome is virus—it's not just you personally. And there are subtle differences in those, so if a taxonomist were to measure viral genes, thinking it was a human gene, they would come up with a very different answer from one that was in the human lineage perhaps from the beginning.

QUESTIONER: You mentioned that there are 100 trillion cells in our body, so to speak. A hundred trillion. Right? Aren't most of them nonhuman? Aren't we really dependent on having a lot of animal cells in our body? And in essence are we not human but a zoo?

VENTER: No. It depends on what you had for breakfast. We have 100 trillion human cells. We have at least that many bacterial cells associated with us. So—

QUESTIONER: So we are a zoo?

VENTER: Well, it depends. There aren't too many bacterial zoos. But an important part of human metabolism is, you're not so much what you eat, as people say; you're what you feed the bacteria in your gut. When we look at the chemicals in the blood after a meal, there are roughly 2,500 compounds that we, as a species, can make. We see roughly twice that many as bacterial metabolites in our guts, from what we feed them. So we live in a bacterial milieu. We breathe it. Our guts, every orifice, our skin—we have more bacterial cells than we have human cells, and they're a key part of our existence. We can't exist in a healthy life without them. So that could be a zoo if you had a microscope.

QUESTIONER: I have two questions, actually: one referring to Richard Dawkins's latest book, *The God Delusion*. I would like to know from Venter how happy was he about this book. Because, well, if there's no God, you can't tinker with Genesis. So, you don't have any ethical problems with that, maybe. The second question is: Do you think humankind is overtaking evolution? So it will happen in the lab, and not in the natural environment anymore?

DAWKINS: The first question seemed rather a strange question. It mentioned my book *The God Delusion*, and then asked, "What's Craig Venter's attitude toward that, because . . ." And I didn't quite understand what the "because" was. Something to do with, "He doesn't have to worry about Genesis anymore." But I don't suppose he ever did worry about Genesis. [Laughter]

VENTER: I guess the assumption is, we can't play God if there *is* no God. [Laughter]

DAWKINS: All the more reason to do so.

QUESTIONER: In response to Mr. Brockman's annual *Edge* question, about where have people changed, where have the two of you changed your minds? And could you comment on Steven Pinker's response to that question, where he stated that he once thought humans were essentially not evolving anymore but now he believes they are?

DAWKINS: Right. The questioner points out that John Brockman's *Edge* website this year has a question, "What have you changed your mind about, and why?" I won't give my answer, because it takes too long to explain. However, I will say that in response to Craig Venter today, I am prepared to change my mind, if he gives a better answer to my question about molecular taxonomy. Maybe now is not the time to do it. I'm on the brink of changing my mind, but I remain highly skeptical as to whether I will in fact have to do so.

VENTER: We'll have to go through some genome data as we follow up on this. I think Pinker thought there was no human evolution because he spent so much time at a university. [Laughter / Applause].

QUESTIONER: We've talked a lot about design and technical things. How about soul? Science tried to figure out where our souls sit. Where is it in your mind, Mr. Venter?

DAWKINS: The question is about where the soul sits. Either the soul doesn't exist at all—and I don't believe it does exist, in the sense of anything outside the brain—or it is a manifestation of brain activity. I certainly would think it highly, highly unlikely that there's anything like a soul that survives the death of the brain. So I think that one of the aspects of the revolution in biology is a complete destruction of dualism and of obscurantist mystification.

QUESTIONER: Craig's comment about being able to, perhaps, in the future take carbon dioxide out of the atmosphere and create fossil fuels—good ones, because we're not digging them out of the ground—is admirable. Of course, it's always dangerous to predict what future technologies will be like. But it seems to me that there are two classes of technological solutions that might be able to use your invention, when you come up with it. One would be some kind of black box that takes the carbon dioxide in the immediate vicinity of the black box and converts it into fuel. And the problem here is that you only have about 400 parts per million of carbon dioxide in the atmosphere. So you would require an enormous amount of processed energy to be able to get enough carbon dioxide to make the quantities of fuel we need.

The other broad class of technological solution would be, perhaps you could create some kind of enzyme—or whatever you would call it—but take advantage of the huge surface area of the oceans, and you could then put it into the ocean, and then it would take carbon dioxide out of the atmosphere and convert itself into oil. But then we would have the problem of the oceans being covered with oil, another undesirable solution.

VENTER: They're thoughtful questions. The first, about the concentration of CO2, is relatively easy to deal with. The KMs of the enzymes and these organisms that exist throughout our planet are able to capture CO2 out of the atmosphere, out of the water. But we don't need to rely on that. We have two phenomenal and, soon a third, point source of carbon dioxide. The two largest are power plants and cement factories. If we could simply capture back the CO2 from those two point sources, it makes it very easy, because of the incredible concentrations you have there, and will eventually get in a cycle of a renewable source from that. We also have a third. It's a clustered carbon dioxide from a variety of sources to be pumped down into oil wells or coal beds. So, we are working in one of our programs with BP, trying to look at converting that CO2 back into methane, so you could constantly be in a recycling mode. Once you sequester CO2, we could use that as a source of energy instead of constantly taking more out of the ground. So we have so many incredible point sources of CO2 production right now that that's the least of our worries.

BROCKMAN: Thank you. Thank you all for coming.

10
The Nature Of Normal Human Variety

Armand Marie Leroi

[March 13, 2005]

Armand Marie Leroi is a professor of evolutionary developmental biology at Imperial College London.

The question that interests me, as it does so many other people, is how to go about making a human being. It's a very difficult problem. Roughly what it boils down to is this: It's often said that the genome is something in the nature of a book. It has words, a grammar, a syntax, and of course those words have meaning. The only problem is that we don't know actually what that meaning is. So the question is, How do we decipher that? And turning that question around, looking at it from the point of view of the human body, what do those genes mean to the construction of the human body?

Of course I don't actually work on humans; they're just too inconvenient. I work on worms. This worm is *Caenorhabditis elegans*, for which Brenner, Sulston, and Horvitz won the Nobel Prize in 2002. And the reason why I and a thousand other scientists work on this worm is that, for all of its marvelous properties, it's easy to keep thousands of them in petri dishes, and it's easy to find mutants. And that's the critical thing. We find mutants that interrupt particular genes, and that tells us what those genes do and what they mean to the body of a worm.

Developmental biologists have been doing this for a long time—once a field has its Nobel, you can be sure it's reasonably mature. What people haven't done, however, is to do this for the human body. The reason is obvious: You just can't go out and generate mutants in humans. For humans, you've got to go out and find those mutants. But they're out there. There are thousands upon thousands of mutants out there—no, more, millions—no, actually billions. This is because we are all mutants. That's one thing you don't expect but which happens to be statistically true. Each of us carries mutations that interrupt particular genes. So if you can just find who is a mutant for a particular gene, and examine what those people look like, you can then work out what those genes do.

This raises a question: What exactly is a mutant? Worm biologists and fly biologists—geneticists generally, working on model organisms—use the word "mutant" in a particular way. In worms and flies, there is an arbitrarily defined strain that we call the "wild type." But in humans there is no arbitrary wild type. So can you, in fact, speak of mutant humans?

You can, but the definition of what is a mutant in humans is necessarily more roundabout, because we have such an extraordinary amount of natural variation in our species. If you go around the world, you see tall people, short people, red-haired people, brown-haired people, people with curly hair, people with no hair, and so on. Given all this variation, what exactly, and who exactly, is a mutant? It's an important question, because to say something is a mutant is to make an invidious distinction. This is to say, something is not just different but actually abnormal in some fashion. Yet despite the fact that there is so much variation in our species, it is possible to speak in a coherent way of mutations and of mutants in humans.

Roughly, the reason you can do so is as follows: If you look at the coding sequence—more precisely, the protein sequence—

produced by any given gene, it's the case that for most genes nearly everybody has the same version. True, there are some genes that are polymorphic—variable—and these are genes that give us our natural diversity. But they actually constitute a very small fraction of the genome. Most people have the same functional version of a gene. Given that fact, you can define a mutant as somebody who has a rare variant of a gene— moreover, a variant that harms him in some fashion. And if you look at it that way, it's clear that we all carry rare variants that do us harm in some way, and that we are in fact all mutants.

We can even put some numbers on this. One of the surprising results in recent years, which comes from the comparison of the genomes of different species, is that every newborn child carries three novel deleterious mutations—that is, mutations that its parents didn't have. Not only that, but each child inherits at least some of the mutations that its parents have as well. It's estimated then—and of course this is just an estimate—that every newly conceived person has something like 300 mutations that affect its health for the worse in some fashion.

Of course, that number doesn't tell us a whole lot. We need to know not only the number of mutations we have but also the distribution of their effects. This is because some mutations have severe effects. They are the mutations that cause the big known inherited diseases—about 10,000 such diseases have been identified so far. But there must be many, many more mutations that do us harm but only subtly so. These are the mutations that give us weak eyes, bad backs, and the like. These are mutations we know very little about but that statistically speaking must be there. It's in these mutations that a lot of human health lies—or rather the absence of human health lies. At least it does once you have got rid of the contagious diseases.

When I speak of mutations that do somebody harm, what I really mean is not so much that they just affect physiological health; what I really mean is that they affect the Darwinian fitness, the probability that they will reproduce. It's an evolutionary definition. It's the kind of definition that can encompass an enormous range of impairments, and the kinds of impairments you see that are caused by mutations are sometimes of a degree and form that you just cannot conceive of.

If you go to teratology museums—literally "monstrosity museums"—in places such as Amsterdam and Philadelphia, you can see rows of babies in bottles. These infants, usually stillborn, are deformed in ways that are truly hideous, that represent the kinds of monstrosities you might expect from Greek myth. I mean this quite literally. They include children born with a single eye in the middle of their forehead, who look exactly like the monsters of Greek myth—Polyphemus in the Odyssey, for example. Indeed, it's sometimes suggested that the monsters of Greek myth were inspired by deformed children, and this seems to be a fairly remarkable correspondence, at least with some of them.

These infants, when you see them, are truly horrific. But very quickly, after you look at them, a sort of intellectual fascination takes over, because it's clear these children tell us something deep about how the human body is built. Take, for instance, the children with a single eye in the middle of their foreheads. The syndrome is called, appropriately, Cyclopia. Cyclopia is caused by a deficiency in a gene called Sonic hedgehog. Sonic hedgehog is named after a fruit-fly gene that, when mutated, causes bristles to sprout all over the fruit-fly larva, hence "hedgehog". When the gene was found in mammals, some wit called it Sonic hedgehog, after the video game character. If you get rid of this gene, bad things happen. You lose your arms beneath the elbow and legs beneath the knee. The face collapses in on itself, such that you get a single eye in the middle of the forehead and the rest of the face collapses into a long, trunklike proboscis. The forebrain, which is normally divided such that we have a left and a right brain—the left and right cerebral hemispheres—is fused into a single unitary structure. Indeed the technical name for this syndrome is called Holoprosencephaly.

Now, all this is horrible, and that's just an initial list of things that can go wrong in infants that have no Sonic hedgehog. But what's really interesting about it is that by looking at infants of this sort, you can reverse-engineer and ask what Sonic hedgehog does in the embryo. Instantly it tells you that one of the things the Sonic hedgehog does is to keep our eyes apart, because if you don't have the gene the face collapses. It also separates the left and and right sides of our brains. And it's needed for the formation of our arms and legs. In fact, it is one of the most ubiquitous and powerful molecules in the making of our bodies.

And other, more subtle mutations tell more about it. For example, just as having too little Sonic hedgehog causes the face to collapse in upon itself, having too much causes it to expand. I was recently in San Francisco, in Jill Helms's lab at the University of California, San Francisco, where she's got a jar containing the head of a pig. Or is it two pigs? It's just not clear, since the jar contains a pig with two faces, two snouts, two tongues, two throats, and three eyes. It's not a Siamese-twin pig; it's just a pig with two faces. Chickens and pigs with two faces crop up periodically, as indeed do humans with two faces, or nearly two faces. There's a syndrome in which you have eyes widely spaced from each other and in which the nose becomes duplicated. You have two noses side by side in two varying degrees of development.

The gene for this syndrome has recently been cloned, and

guess what? It turns out to be the gene that controls Sonic hedgehog and in fact switches it off. People with the syndrome have too much Sonic hedgehog, just as infants with Cyclopia have too little. So by looking at a range of these kinds of syndromes, you can put together a complete picture of how a gene like Sonic hedgehog controls one particular feature of us, the width of our faces. It's a mundane thing that you'd hardly think about but that seems to be controlled by this genetic system.

There are many other disorders that are equally informative. The star at the Mütter Museum at the College of Physicians of Philadelphia is Harry Eastlack, a man who had a disease called fibrodysplasia ossificans progressiva. It's a disorder in which supernumerary bones form. The Mütter Museum has his skeleton, which he donated at the time of his death when he was in his forties. The skeleton is essentially not one man's skeleton—it is, as it were, one skeleton encased in another. What happens in this disorder is that wherever you get a bruise or a wound, instead of normal cells moving in to regenerate the skin and the flesh and heal the wound, bone forms. So every bruise turns to bone. The children are born relatively normal, but as they go through life, bone accretes all over them, such that they can no longer move. They become rigid, locked into place. You can cut it away, of course, but as soon as you make an incision and that incision heals, more bone forms. So it's a vicious circle. We don't know which gene is mutated in this syndrome. But it's almost certainly got something to do with bone morphogenetic protein, a protein that is, as the name suggests, normally involved in making bone in infants. It's just that in most of us this gene switches off. In these people, this protein keeps on being produced throughout life, especially when there's a wound. It's another marvelous instance of how a given mutation can tell us something important about how bones are formed. FOP is a very rare disorder, and the reason the gene hasn't been cloned is because to identify genes, to clone genes, you need to have big pedigrees—at least, it helps. But these people just never have children.

People sometimes ask what developmental biology is good for. We can identify genes that are responsible for making this or that part of the human body. But in humans, of course, there's a pressing question: namely, how can you fix the deleterious consequences of these mutations? It's one thing to go into a clinical genetics ward, a pediatric ward, and study kids who are seriously deformed, and say, "This is just terribly interesting. Your son is highly informative about the function of the Jagged-2 gene." But that's not a great deal of comfort to the parents, who actually have to deal with raising children who are variously deformed and may die or, at the very least, have to undergo a great deal of surgery. That, of course, is the problem. The molecular biology is beautiful, but when it actually comes to curing people, you just have surgery—which is little more than a rather sophisticated form of butchery.

The great promise, of course—and it's been a promise for years now and will remain so for some time—is that by learning about what genes do and how organs and tissues are constructed we can reconstruct them as we wish. By working out the program, we can take cells, put them in a test-tube, and rebuild tissues. You don't have cartilage in your larynx? We can build it for your child, and we can fix it. You don't have a breast? We can rebuild that, too. And so on. This is a whole new area, called tissue engineering. There are big institutes devoted to it now, where engineers, materials scientists, and molecular biologists are all working together. So far, it must be said, it's more institution-building and propaganda than real results, but it'll happen. That is when the justification for this whole science will ultimately come.

There's no doubt that when you see some of these children who are so terribly deformed, it's very difficult. It's shocking, it's heartbreaking, and if you spend any time with them, whether they're alive in pediatric wards or they're just babies in bottles, it takes a real psychological toll. I certainly don't ever get completely hardened to it. But what is also true is that the intellectual fascination of seeing what has gone wrong in these unusual bodies takes over. This is especially the case when your eye is attuned to perceiving the differences in detail, once you see that it's not just arbitrary deformity, once you understand that what you're really looking at are the outcomes of the laws that regulate and make the human body. When this happens, deformity acquires a real beauty. It's a beauty that emerges from answering one of the oldest questions in biology: namely, how are we put together?

But that's intellectual beauty. What of human physical beauty? This is something that interests me greatly. I'm not interested in the general aesthetic question here, but in ourselves. Some people say that beauty is uninteresting and just a matter of taste. I don't think so. I would say, and there are others who would agree with me, that we have a general psychological program from which stems a universal notion of beauty. Incidentally, this idea that we all perceive certain features to be beautiful is one that Darwin would have disagreed with. Darwin believed that the perception of beauty was particular to particular peoples in particular times and places. He was probably wrong, or at least he was only partly right. I won't attempt to justify that answer, but I think it to be true. These days, the general thinking tends to be that there's a universal notion of beauty which is true for people around the world. And the question is, What is that and what drives it?

Many people think that beauty is a certificate of health; this is an idea that comes out of sociobiology. But it is more obvious than that. It's simply the idea that beautiful people are healthy people and we search for healthy mates. And that's probably true. Or at least it was. But is it still? In the past, health was primarily a matter of environmental conditions—your exposure to contagious diseases and the amount of food you had when you were growing up. Rich people had better environments, hence the positive association between beauty and wealth. But what of modern economically egalitarian societies, such as Holland? In such societies, does the ancient association still obtain? If the variance in beauty is due to the variance in the quality of the rearing environment, then it must be the case that the Dutch—who all eat much the same good food, live in much the same well-designed houses, and have access to much the same excellent health care—must all be equivalently beautiful. But is this so? The answer is, of course, no. Among the Dutch, you can find good-looking and not so good-looking people. And the question is then, why?

I would argue that the reason for this is that there is and will always be variance in beauty, because there is variance in mutational load. What is beauty fundamentally about? I would argue—and this is really just a postulate at this time but it is one that interests me a great deal—that the fundamental reason some of us are more beautiful than others is because of those deleterious mutations we all carry We may carry 300 deleterious mutations on average, but there is of course a variance associated with that. Not everybody has 300. Some people have more, some people have fewer. If this is true—and statistically it must be true—then someone in the world has the fewest mutations of all. Someone in the world is the least mutant human of all. Indeed, we can actually calculate, making some assumptions about the shape of the distribution, how many mutations that person has—and it turns out to be 191 versus the average of 300. This, to my mind, is surprisingly many. I would suggest that if we could find that person, he or she would be a good candidate for being the most beautiful person in the world. At least she would be, assuming she did not grow up in some impoverished underdeveloped nation. Which, statistically, she will have done, since most people do.

There's one more thing I should like to know about, and that is the nature of normal human variety. There are tens of thousands of geneticists around the world, all of whom are busy identifying the genes that cause human disorders of one sort or another. Historically they began with the really easy ones, the big congenital disorders, especially those that allow people to survive and produce children and so have big pedigrees that allow mapping of the genes. Now the emphasis has shifted to studying the genetic basis of more subtle and more complicated kinds of disorders—things like diabetes and cancer—that have lots of genes that underlie them, each of which has a small effect. This is a much more difficult task, but people are doing it because these are the inherited diseases that affect millions.

But there's one aspect of human inheritance that people are resolutely ignoring. And that is normal human variety. Or, to put it more crisply: race. If we look around the world, we find that people look very different from each other. These differences are manifestly genetic. They must be. That's why people's kids look like them. Yet we know nothing about that variety. We don't know what the differences are between white skin and black skin, European skin versus African skin. What I mean is, we don't know what the genetic basis of that is. This is amazing. I mean, here's a trait, trivial as it may be, about which wars have been fought, which is one of the great fault lines in society, around which people construct their identities as nothing else. And yet we haven't the foggiest idea what the genetic basis of this is. Why is that?

The reason is twofold. The first—which is not such a trivial problem—is that skin color is not controlled by one gene. If it were only one gene, we'd know it. It's many genes—more than three but certainly fewer than 30. It's a difficult problem, although, frankly, it's not such a difficult problem that if geneticists really wanted to solve it, they couldn't. It'd be easy enough to do if they put a fraction of the effort that went into discovering the BRCA1, the breast cancer gene. I'm not saying they should; finding the breast cancer gene is more important than discovering the genetic basis of white and black skin, but still, it's not a technically impossible thing to do.

But of course the fundamental reason why people don't do it is because it's race genetics. It's because of the long and sorry history of genetics and racial differences. And indeed, more than that, the whole thrust of genetics since the war has been to argue that races don't exist and that they are just social constructs. This is very much the Harvard School—Dick Lewontin, for example, has been one of the big proponents of this point of view. The late Stephen Jay Gould was another.

After World War II, when the enormities of Nazi science really hit home—which were in turn the consequence of a much larger racial science not just in Germany but everywhere—all right-thinking scientists made a resolute effort to ensure that science would not be bent to such evil purposes again. They were determined that science would never again be used to make invidious discriminations among people. The immediate result of this was the UNESCO Declaration on Race in 1950—fronted by Ashley Montagu and backed up by geneticists such as Theodosius Dobzhansky—which affirmed the equality of races. Then, in the 1960s, Dick Lewontin and others discovered that gel electrophoresis could be used to survey genetic variation among proteins. These studies showed that humans have a huge amount of concealed genetic variation. What is more, most of that genetic variation existed within continents or even countries rather than among them. UNESCO said races were equal; the new genetics said they didn't exist. Finally, moving a few decades on, the Out-of-Africa hypothesis of the origin of *Homo sapiens* comes to the fore and multi-regionalism falls from fashion, as it becomes clear that humans are not only a single species—something we've known since Linnaeus' day—but a single species that has diverged into sub-populations only very recently.

The result of this history—which has been driven partly by data and partly by ideology—is that these days anthropologists and geneticists overwhelmingly emphasize the similarities among people from different parts of the world at the expense of the differences. From a political point of view, I have no doubt that's a fine thing. But I suggest that it's time we grew up. I would like to suggest that by emphasizing the similarities but ignoring the differences we are turning away from one of the most beautiful problems that modern biology has left: namely, what is the genetic basis of the normal variety of differences between us? What gives a Han Chinese child the curve of her eye? The curve I read once described by an eminent Sinologist as the purest of all curves. What is the source of that curve? And what gives a Solomon Islander his black-verging-on-purple skin? Or what makes red hair?

Actually, the last is the one thing we do know. It turns out that red hair is due to a mutation in a gene called MC1R, melanocortin receptor 1, which controls the production of eumelanin, black pigment, versus red pigment, pheomelanin. Rather marvelously, it also turns out that mutations in MC1R also cause red hair in red setters, Scottish cattle, and red foxes. But we don't know what causes brown eyes versus blue eyes versus green eyes. We know very little about the variation in normal human height. We don't know why some girls have big breasts and some have small breasts. These are important questions—or at least jolly interesting ones—and we just don't know their answers.

The reason I love the problem of normal human variety is because, almost uniquely among modern scientific problems, it is a problem we can apprehend as we walk down the street. We live in an age when the deepest scientific problems are buried away from our immediate perception. They concern the origin of the universe. They concern the relationships of subatomic particles. They concern the nature and structure of the human genome. Nobody can see these things without large bits of expensive equipment. But when I consider the problem of human variety, I feel as Aristotle must have felt when he walked down to the shore at Lesvos for the first time. The world is new again. What is more, it is a problem we can now solve, a question we can now answer. And I think we should.

Of course, there will be people who object. There will be people who will say that this is a revival of racial science. Perhaps so. I would argue, however, that even if this is a revival of racial science, we should engage in it, for it does not follow that it is a revival of *racist* science. Indeed, I would argue that it is just the opposite. How shall I put it? If you want to prove, what most of us believe, that skin color does not give the measure of a man, that it tells nothing about his abilities or temperament—then surely the best way is to learn about the genetics of skin color and the genetics of cognitive ability and demonstrate that they have nothing to do with each other. The point is that there will always be people who wish to construct socially unjust theories about racial differences. And though it is true that science can be bent to evil ends, it is more often the case that injustice creeps in through the cracks of our ignorance rather than through anything else. It is to finally close off those cracks that we should be studying the genetic basis of human variety.

11
Brains Plus Brawn

Daniel Lieberman

[October 17, 2012]

Daniel Lieberman is Lerner Professor of Biological Sciences and a professor of human evolutionary biology at Harvard University.

I've been thinking a lot about the concept of whether or not human evolution is a story of brains over brawn. I study the evolution of the human body and how and why the human body is the way it is, and I've worked a lot on both ends of the body. I'm very interested in feet and barefoot running and how our feet function, but I've also written and thought a lot about how and why our heads are the way they are. The more I study feet and heads, the more I realize that what's in the middle also matters, and that we have this strange idea—it goes back to mythology—that human evolution is primarily a story about brains, about intelligence, about technology, triumphing over brawn.

Think about Greek myths like the myth of Prometheus and Epimetheus. Epimetheus, which means "hindsight," is the Titan who gave out all the gifts to the animals, and when he finished, he hadn't given any gift to humans. Prometheus took pity on these poor humans who didn't have claws and fangs and speed and power, so he gave humans fire. Of course, he got tortured by the rest of the gods for this. I think this idea—that humans are essentially weak creatures—is deeply woven into a lot of the ways we think about our bodies.

Another good example would be the Piltdown hoax. The Piltdown forgery was a fossil that was discovered in the early 1900s, in a pit in southern England. This fossil consisted of a modern human skull that had been stained and made to look really old, and an orangutan jaw whose teeth had been filed down and broken up—all this thrown into a pit with a bunch of fake stone tools. It was exactly what Edwardian scientists were looking for, because it was an apelike face with a big human brain, and also it evolved in England, so it proved that humans had evolved in England, which of course made sense to any Victorian or Edwardian. It also fit with the prevailing idea at the time of G. Elliot Smith, that brains led the way in human evolution—because, if you think about what makes us so different from other creatures, people always thought it was our brains. We have these big, enormous, fantastic brains that enable us to invent railways and income tax and insurance companies and all those other wonderful inventions that made the Industrial Revolution work.

It turned out that the Piltdown Man was a fraud, and it turns out that the idea that brains got large early in human evolution is incorrect as well. We now know that humans and chimpanzees split maybe around 6 million to 7 million years ago, and the very earliest hominins—creatures that are more closely related to humans than to chimpanzees—had really small brains. In fact, early *Australopithecus,* like Lucy, also had quite small brains. Even the early members of the genus *Homo* had small brains.

Tools first started appearing around 2.6 million years ago, and those hominins have slightly larger brains than *Australopithecus.* But if you factor out the effects of body size, what's called their encephalization quotient—the ratio of brain size to body size—for what you expect for a mammal of a body size versus what you got, it was not that much bigger than chimpanzees or early Australopiths. To put it in perspective, an EQ of 1 means that your brain size is exactly the size of a brain you predict for your body size. Chimpanzees have an EQ of 2.1, and humans have an EQ of about 5.1. Australopiths have EQs of about 2.5, and the earliest members of the genus *Homo* have EQs of about 3.0 to 3.3. Their brains are a little bigger than a chimpanzee's but not hugely so, and it wasn't until long after the genus *Homo* evolved that brains started getting really large. So increases in brain size were not an early event in human evolution, and in fact they didn't occur until after the invention of hunting and gathering—not until cooking and various other technological inventions that gave us the energy necessary to have really large brains.

Brains are costly. Right now, just sitting here, my brain—even though I'm not doing much other than talking—is consuming about 20-25 percent of my resting metabolic rate. That's an enormous amount of energy, and to pay for it I need to eat quite a lot of calories a day, maybe about 600 calories a day, which, back in the Paleolithic, was a difficult amount of energy to acquire. So having a brain of 1,400 cubic centimeters, about the size of my brain, is a fairly recent event and very costly.

The idea then is, At what point did our brains become so important that we got the idea that brain size and intelligence mattered more than our bodies? I contend that the answer was "Never", and certainly not until the Industrial Revolution.

Why did brains get so big? There are a number of obvious reasons. One of them, of course, is for culture and for cooperation and language and various other means by which we can interact with each other, and certainly those are enormous advantages. If you think about other early humans, like Neanderthals, their brains are as large or even larger than the typical brain size of human beings today. Surely those brains are so costly that there would have had to be a strong benefit to outweigh the costs. So cognition and intelligence and language and all of those important tasks that we do must have been very important.

We mustn't forget that those individuals were also hunter-gatherers. They worked extremely hard every day to get a living. A typical hunter-gatherer has to walk between 9 and 15 kilometers a day. A typical female might walk 9 kilometers a day, a typical male hunter-gatherer might walk 15 kilometers a day, and that's every single day. That's day-in, day-out; there's no weekend, there's no retirement, and you do that for your whole life. It's about the distance if you were to walk from Washington, DC, to LA every year. That's how much walking hunter-gatherers did every single year.

In addition, they were constantly digging, climbing trees, using their bodies intensely. I would argue that cognition was an extremely important factor in human evolution, along with language, theory of mind—all those cognitive developments that make us so sophisticated. But they weren't a triumph of cognition over brute force; it was a combination. It was not brains over brawn, it was brains *plus* brawn, and that made possible the hunter-gatherer way of life.

What hunter-gatherers really do is they use division of labor. They have intense cooperation, they have intense social interactions, and they have group memory. All of those behaviors enable hunter-gatherers to interact in ways such that they can increase the rate at which they can acquire energy and have offspring at a higher rate than chimpanzees. It's an energetically intensive way of life, made possible by a combination of extraordinary intelligence, inventiveness, creativity, language, but also daily physical exercise.

The other reason we often discount the importance of brawn

in our lives is that we have a strange idea of what constitutes athleticism. Think about the events we care about most in the Olympics. They're the power sports—the 100-meter dash, the 100-meter freestyle events. Most athletes, the ones we value the most, are physically very powerful. But if you think about it this way, most humans are wimps.

Usain Bolt, who is the world's fastest human being today, can run about 10.4 meters a second, and he can do so for about 10 or 20 seconds. My dog, any goat, any sheep I can study in my lab, can run about twice as fast as Usain Bolt without any training, without any practice, any special technology, any drugs, or whatever. The fastest human beings are incredibly slow compared with most mammals—not only in terms of brute speed but also in terms of how long they can go at a given speed. Usain Bolt can go 10.4 meters a second for 10 to 20 seconds. My dog, or a goat, or a lion, or a gazelle, or some antelope in Africa, can run 20 meters a second for about 4 minutes. There's no way Usain Bolt could outrun a lion, or, for that matter, run down any animal.

A typical chimpanzee is between about two and five times more powerful than a human being. A chimpanzee who weighs less than a human can rip somebody's arm off or rip their face off (as recently happened in Connecticut). It's not that the chimpanzee is remarkably strong, it's that we are remarkably weak. We have this notion that humans are terrible natural athletes. But we've been looking at the wrong kind of athleticism. What we're really good at is not power; what we're really phenomenal at is endurance. We're the tortoises, not the hares, of the animal world. Humans *can* actually outrun most animals over very, very long distances.

The marathon, of course, is an interesting example. A lot of people think marathons are extraordinary, and they wonder how many people can run them. At least a million people run a marathon every year. If you watch any major marathon, you realize that most of those folks aren't extraordinary athletes, they're just average moms and dads. A lot of them are charity runners who decided to raise money for some cause, cancer or diabetes or something. I think that proves that your average human being can run 26.2 miles without that much training, or much ability to be a great athlete. Of course, to run a marathon at fast speeds is remarkable, but again, it just takes some practice and training; it's not something extraordinary.

We're remarkable endurance athletes, and that endurance athleticism is deeply woven into our bodies, literally from our heads to our toes. We have adaptations in our feet, our legs, our hips and pelvises, and our heads and our brains and our respiratory systems. We even have neurobiological adaptations that give us a runner's high, all of which help make us extraordinary endurance athletes. We've lost sight of just how good we are at endurance athleticism, and that's led to a perverse idea that humans aren't very good athletes.

A good example are those races where humans are compared with horses. In Wales, this started a few years ago. I guess it started out as a typical drunken pub bet: Some guy bet that a human couldn't beat a horse in a marathon. They've been running a marathon in Wales for the last, I think, 15-20 years. To be fair, most years the horses beat the humans, but the humans often come very close. Whenever it's hot, the humans beat the horses. They also now have ultra-marathons in Arizona where humans race horses. Again, most years the horses beat the humans, but every once in a while, the humans do beat the horses. The point is not that humans are poor athletes because the horses occasionally beat us, but humans can actually compete with, and often beat, horses at endurance races. Most people are surprised at that.

These are horses with riders on them. One of the interesting things about these races also is that they're so worried about the horses getting injured that the horses have mandatory veterinary check-ups every 20 kilometers, but not the humans, because humans can easily run 40 kilometers without injury. But if you make a horse gallop for more than 20 kilometers, you seriously risk doing long-term permanent musculoskeletal damage to the horse.

Until recently, you couldn't live, you couldn't survive as a human being without being an endurance athlete. Not just hunting and gathering required athleticism but also being a farmer. Subsistence farmers had to work extremely hard. Until the invention of industrialized machinery, farmers had to work even harder than hunter-gatherers, often spending many thousands of calories a day. They had to dig ditches and throw vast quantities of hay into bales and they had to schlep stuff all over the place. Farmers had to work brutally difficult, hard, exhausting lives. It wasn't until, again, the invention of new technologies, such as domesticating animals or, even more recently, machinery such as the internal combustion engine, that farmers were able to live non-grueling lives.

It's only in the last 100 years—in fact, maybe for many people the last 20, 30, 40, 50 years—that human beings have been able to lead sedentary lives without any physical activity. A typical American today, living in a post-industrial information-technology world, can get up in the morning, reach into the cabinet, get breakfast cereal that's in a box, pour it, reach into the refrigerator, pour some milk. You can spend your entire day without ever elevating your heart rate. You can sit in wheelie chairs all day long, take an elevator to your floor, sit in a chair all day long staring into a computer, and drive home from work. Dinner basically involves pressing a few buttons, or if you have to go to the supermarket, wheeling around a cart. Human beings today don't have to engage in physical activity anymore. It's natural for us to think the world around us is normal. But our world, our lives today, are profoundly abnormal, especially in the case of no longer having to use physical activity as part of our daily routine.

We've also got this bizarre notion, which finally came true, that our bodies don't really matter. Think about St. Thomas Aquinas, who believed that the flesh was irrelevant and all that really mattered was your soul. Only monks and rich clerics in Europe could get away with that kind of opinion in the Middle Ages, because everybody else had to work like a dog. But now your average human being can live a life of ease and luxury, much more physically inactive than even kings and queens of Europe in the old days. We've finally been able to realize Aquinas's dream of basically ignoring the flesh, ignoring the body. The result is an epidemic of obesity and heart disease and various kinds of cancers, such as colon cancer and breast cancer, that largely come from physical inactivity combined with a really loathsome diet.

What's interesting is that there's a reversal of class today. Now, it's only the very wealthy who can afford to be physically active and eat a good diet, and it's the lower classes, the people in the working class, who cannot afford to be physically active. People who are well off, the 1-percenters, or maybe the 5-percenters, of our country, can afford to go to health clubs and take yoga classes and buy organic food. But most Americans today, particularly those in the working class, have to work all day long, have to commute long hours. They don't have time to exercise; they can't afford to buy food that's healthy. The result is that there's now an interesting reversal that never occurred before in human evolution, which is that the wealthier you are, the healthier you are. It used to be that it was only the very wealthy who could afford to get heart disease.

The first example we know of heart disease is actually from a CT scan of an Egyptian princess from her sarcophagus. She was mummified, and a CT showed that she's the oldest case of heart disease. It's not coincidental that this was a pharaoh's daughter. She clearly was able to hang around in the palace all day long and eat too much and not have to exercise. But that was a luxury, until recently, of only the wealthy, and now it's completely flipped.

I'm a professor of human evolutionary biology, and I've long been fascinated by heads, because our heads are really unusual. If you were to meet a Neanderthal or *Homo erectus* on the street in New York City, for example, you'd probably stare mostly at their heads. Their bodies would be very similar to yours and mine, but they'd be really different from the neck up. If there's one important part of the human body we need to understand in order to understand why we're the way we are, it's our heads.

Heads are complicated. If you think about your head, almost every particle that enters your body goes through your head. Everything you breathe, everything you eat, and everything you smell. Almost everything goes through your head, and a lot of major functions also go on in your head. You speak in your head, you smell, you taste, you chew, you think, your sense of balance is in your head—I could go on. And yet all of that is packaged into a tiny space about the size of a soccer ball, and if anything goes wrong with anything, you're dead. If you can't breathe, speak, swallow, smell, taste, hear properly, until recently you were immediately selected out of the gene pool.

Heads are also fascinating. You'd think because of all that complexity that they'd have to be the most constrained part of the body, right? You can't let anything go wrong in a head or you're dead. But it turns out that heads are pretty much the most evolvable parts of our bodies. They're the parts of our bodies that seem to have changed the most. As I said, if you met a Neanderthal or a *Homo erectus* you'd be mostly different from the neck up, not from the neck down.

I've long been interested in how heads can be so evolvable and what the story of the head tells us about human evolution. Part of that is about the brain. Human brains got bigger and that's a major effect on heads, but your brain is actually the same size as the brain of a Neanderthal. Actually, it's a little smaller than a Neanderthal's, and our brains are only a little larger than that of a late *Homo erectus*. What's different about our heads is not the size of our brains.

The most important difference is the size of our face. Humans are profoundly different from every other hominin. Our difference from other hominins is primarily the size and position of our face. What we've done in our recent evolution is we've shrunk our face and retracted it. Our face is now underneath our brains rather than sticking out in front of our brains. That's why we don't have big brow ridges, and that's why our tongues and mouths are small, which causes our larynx to be low and changes the shape of the vocal tract.

In working on heads, I came to the conclusion that although brains are really important in human evolution, there's much else about the head that's unique, that doesn't have that much to do with the brain.

I'll give you an example related to running. When we run, we can't control our heads the way any other animal can. If you watch a dog or a horse run, you'll see that its head is like a missile on a body. The body is moving and the head stays stock-still. That's for an important reason: because of gaze. You need to stabilize your gaze in order to see where you're going. We need to have a stable image in order for us to evaluate it, to use that information. If you're running across a landscape and the world is jiggling horribly, you won't be able to see where rocks are, you won't be able to see where other obstructions are, you won't be able to see where your prey is. You won't be able to function effectively. For example, if you were to trip over a rock in the Ice Age, that was a huge selective problem. Today if we trip and sprain our ankle or break our leg—OK, it's a big pain, but you can go to the doctor and they'll set it straight and you can use crutches. But imagine that you sprained your ankle 15 miles from home in the Pleistocene. You'd be easy pickings for a saber-toothed tiger or a lion or some other animal. Injuries, damage to your body, was a far greater selective problem in human evolution than it is today.

The reason a dog, or a horse, or any other quadruped, can keep its head still is because its neck is cantilevered off its thorax, so it sticks out horizontally, and the head attaches to the neck from the back, and the neck is also sticking out horizontally. The three units—the head, the neck, and the body—can all rotate independently, thereby keeping the head still. That's what an animal does when it runs. Humans have a little neck, which emerges from the center of the base of our skull, and it's short. We're basically like pogo sticks. By becoming bipeds, we have lost all those mechanisms available to quadrupeds to keep their heads still.

But it turns out that we've evolved special mechanisms to keep our heads still. One of them—the semicircular canals, the vestibular system in our heads—is especially enlarged and gives us enormous sensitivity to pitching motions. The semicircular canals are organs of balance, which essentially function as an accelerometer. As your head pitches forward, which it does every time you hit the ground when you run, the enlarged semicircular canals are sensitive to these angular accelerations. Through a three-neuron circuit to our brain, this activates, without any conscious effort, the eye muscles, which then stabilize the gaze. So, even when your eyes are closed and you move your head, your eyes and the semicircular canals, through that three-neuron system, operate those muscles, keeping your gaze stabilized. It's that fundamental a system.

Even more interesting—we've harnessed our arms and our butts to stabilize the head. When you run, your head wants to pitch forward, which explains why we've lost a lot of the musculature of the upper body. If you look at a chimpanzee, it has this huge set of muscles that connect the head to the shoulder; the trapezius muscle is gigantic in a chimpanzee. They have muscles we don't have, like the atlanto-clavicularis. There's another muscle called the rhomboid, which in humans is a little muscle that goes from the scapula to the spine. In a chimpanzee, it actually inserts on the head. The reason the chimpanzee has all those muscles going from the shoulder to the head is that it enables the animal to climb effectively.

We gave up climbing. We're the worst tree climbers in the primate world. We're actually bizarre as primates, not being very good in trees. Probably almost none of my audience has been in a tree today, which is bizarre for a primate. The reason we gave up climbing was not because of our walking but because of our running.

It turns out that we use our arms to stabilize our heads while running. When you run and you pump your hands opposite your legs, each arm has a mass that's about the same as your head's, and the inertial force that causes your head to pitch forward also causes your arm to fall—that is, the trailing arm. We have a special muscle called the cleidocranial trapezius, which is a slip of a muscle, about the thickness of a pencil. It goes from the clavicle and inserts on a midline structure in the head, which is in the sagittal plane, the midline structure. It turns on just before your foot hits the ground. That muscle then acts as a mechanical strut between the mass of the arm, which is falling, and the head, which is falling forward. And it connects that mass to a spring-like structure called the nuchal ligament, which is aligned in the midline of the head. And so the arm essentially pulls back your head, just as it wants to pitch forward. We thus have what's called a passive mass damper system. It works without your ever having to know it. All you have to have is a pattern generator, a muscle that turns on automatically before your foot hits the ground; and the body automatically corrects for a problem that arose because we're bipeds. It's evidence that humans evolved for running—that we started running maybe 2 million years ago. It's evidence that running is a fundamentally important part of our biology. It's evidence that athleticism is a fundamentally important part of our biology.

I started looking at other features in the head which relate to endurance. For example, we have enlarged noses. No other primate can pick its nose. We have this large proboscis, with this extra vestibule in the front. Why do we have that? It turns out that that vestibule, that extra enlarged portion up there, is a turbulence generator. Air has to go upward through a little valve, called a venturi throat, which creates turbulence as the air then goes into the nose. Then it has to turn at a right angle, which creates more turbulence. Then it goes through another little venturi valve to get into the middle part of the nose, where all the business of the nose is, where all the mucous membranes are that exchange heat and exchange moisture. Because of turbulent airflow inside the nose, there's no longer what's called laminar flow in the nose. Air doesn't just stream into our noses; it's highly vorticial; and by being vorticial, it slows the flow rate down and this means there's no longer a boundary between the air running through the nose and the mucous membranes of the epithelium in there. There's intimate, prolonged contact between air from the outside world entering the nose and the mucous membranes. This enables us to be extremely efficient at humidifying and warming air as it comes in, and extremely efficient at capturing that humidity on the way out, so we don't dehydrate.

Many other features in the head help us to become exceptional long-distance walkers and runners. I became obsessed with the idea that humans evolved to run long distances, evolved to walk long distances, basically evolved to use our bodies as athletes. These traces are there in our heads, along with those brains.

I became interested in the relationship between heads and feet and running and athleticism and human evolution because of my research, but also because of that kind of interaction between research and life. I love to run. I started running when I was a teenager. I was never good at anything, never picked for any team. I was never a track athlete; I never ran for anybody; I just did it because it made me feel good. I figured out in high school that if I didn't run a few times a week, I would go nuts. I started by becoming just a jogger, basically. I ran in college, and I ran in graduate school, and I ran when I was junior faculty. Just a few times a week, a few miles, just to kind of keep sane.

As I started studying the evolution of running and the features in our heads that made us good at running, I started thinking more about my own running. Before I knew it, this led to an interesting feedback relationship between my research and my pastime. We were doing some experiments, trying to figure out how the arm stabilizes the head, and in the lab we were trying to figure out ways to take the arm out of the equation. We would put people in straitjackets while running on treadmills, or have them hold cups of water, whatever we could do. I remember I was running through some park in Cambridge, Massachusetts, and I was probably doing something weird with my arms, holding them above my head, and I remember hearing some guy say, "Oh, that guy clearly has no idea how to run properly," which I thought was funny. But I also realized it was probably true. Although I was really interested in running, loved running, I actually wasn't a very good runner.

That's when we started studying barefoot running. I published a paper in *Nature* with a colleague of mine, Dennis Bramble, in 2004 ["Endurance running and the evolution of *Homo*," *Nature*, 432, 345–52 (2004)] on the evolution of running. We made the case, which started with this work on the head, that humans evolved to run long distances and that the traces of our evolutionary history as runners started about 2 million years ago. Running was important for the evolution of hunting. It enabled early humans to hunt, and that helped release a constraint on brain size, and it wasn't until after hunting and after running that human brain size started to increase.

Not long before the 2005 Boston marathon, a Nor'easter came into town. It was unbelievable. The rain was coming horizontally and people were worried about how they were going to run the marathon in this kind of weather. (By the way, I have run a marathon in a Nor'easter, so I can tell you how horrible it is.)

I gave a big public lecture about the evolution of running and why we run the Boston marathon. There was a guy sitting in the front row; he had a big beard and he had on suspenders but what was most interesting about him was that he was wearing socks wrapped in duct tape. I remember thinking it was some homeless guy from Harvard Square who had come in out of the storm—who would just do anything to get out of the rain. But it turns out he was a Harvard grad who lived in Jamaica Plain and had a bicycle shop. He came up to me afterward and said, "You know, I love running, and I hate wearing shoes, and I'm a barefoot runner. In fact, I just don't like shoes. Humans obviously evolved to run barefoot. Am I weird or am I normal?" I thought, "What a great question!"

We know almost nothing about barefoot running. Obviously, humans did run barefoot for millions of years, and he must be normal from an evolutionary perspective, whereas people like me, who wear shoes, are abnormal. At that time, I was struggling with plantar fasciitis. You get up in the morning and you have this pain in your foot, in the first few steps, because the plantar fascia, a sheath of connective tissue, gets stretched and inflamed and it's poorly vascularized, and it really has a hard time healing. I was buying new running shoes every 250 miles, which is expensive. I thought, "We should study this guy."

I got his e-mail address, brought him to the lab, and he ran barefoot in the lab. When we ran him across the force plate, he ran in this perfectly beautiful, light, and gentle way. Most Americans, when we run we land on our heels. We wear these big, thick, cushioned running shoes with lots of support and lots of cushioning, and they make it really comfortable and let people slam onto the ground on their heels. But this guy didn't land on his heels. He landed with what's called a forefoot strike. He landed on the ball of his foot and then his heel touched down. He had no impact peak.

An impact peak is a collision force when you have an exchange of momentum between two bodies. Momentum is mass times velocity, and the period over which the body comes to a dead stop is when you exchange that momentum. When you drop anything really heavy onto the ground, there's a big peak of force. You hear it when you drop something onto the ground. It had long been known that when people forefoot strike, there's no big peak of force. It's a gentle landing. But nobody bothered paying much attention to it, because most people land on their heels, not on their forefoot.

In fact when we were doing experiments on head stabilization, I used to hate the forefoot strikers who came into the lab, because their heads didn't jiggle as much, since they were landing lightly and gently. It occurred to me when this guy, barefoot Jeffrey, ran across our force plate, that he must be normal and I must be abnormal in the way I'm running in my stupid running shoes.

We started getting other barefoot runners into the lab and found that all of them ran that way. If you were to take your shoes off and run down Fifth Avenue, or Mass Ave., or wherever you live, you'd quickly stop landing on your heel, because it's profoundly painful. You can't slam into the ground with each step. You quickly transition into landing on the forefoot, because it doesn't have an impact peak. So we did some research figuring out what the mechanics of that were and what the physics was, and it turned out to be an interesting story.

We also went to Africa and looked at people who had never worn shoes in their lives. It made me realize that running is a skill, and that we blunt that skill with technology. We wear these fancy running shoes, we drink Gatorade, we do all this stuff, and we no longer really pay attention to how our bodies are functioning. We don't really run very well. The world's best runners all grew up barefoot, and they're all phenomenal runners.

Recently, over the last five or ten years, I've been paying much more attention not just to my own running and how I use my body but also to how people outside the Western world use their bodies—what those skills are and what that can teach us about how we can use our bodies so that we don't get injured.

What I'm doing is part of a gradual, barely growing, movement in biology and evolutionary biology—which we hope will become part of a larger movement in science in general. And that is, using evolution to give us insights into how we use our bodies. And medicine. There's a growing field of evolutionary medicine, started by George C. Williams in the 1990s. He and Randy Nesse wrote a really important book called *Why We Get Sick*, which helped launch the field of evolutionary medicine.

Like many evolutionary biologists, I was inspired by *Why We Get Sick*. In fact, I got fascinated by the question of how the study of the evolution of the human head, the study of the evolution of running and the evolution of athleticism, matters to human health today. There's a disconnect between evolutionary biology and not just the general public but also people in other branches of biology, especially medicine, who should be interested in evolution but aren't. They still think evolution is irrelevant to medicine; they think evolution is irrelevant to much of biology. In fact, if you put the word "evolution" in an NIH grant proposal, it's probably the fastest way to not get funded. It's just not considered important.

The work we did on the evolution of running touched a nerve. I got more than 1,000 e-mails after the first "Born to Run" paper in 2004, and recently, with the barefoot-running paper, I can't even tell you how many e-mails I've got. I still get five to ten e-mails a day from barefoot runners and shod runners all over the world. They're really interested in the evolution of running—not just because it explains why they love to run but also because understanding the evolution of running and of walking helps explain why it's so important to human health. Even more important, from my perspective it also tells us something about how we should be using our bodies.

Studying barefoot running helps us to understand how the body was evolved to run. We didn't evolve to wear cushioned shoes and crash into the ground. We evolved to run lightly and gently, because it hurts to land on the ground the way people do in shoes. Those adaptations, those sensory proprioceptive adaptations in the foot that cause pain, are probably just that—adaptations.

It also occurred to me that so much that's been done in evolutionary medicine has been primarily in infectious disease and in reproduction. We think a lot about the evolution of diseases like tuberculosis and swine flu and avian flu. Those are all evolutionary problems that have direct, immediate relevance to human beings. We spend a lot of time in evolutionary medicine thinking about reproduction and the conflict between parents and offspring, and the placenta, and energy, and nutrition. But evolutionary medicine is also relevant to many other aspects of our body, including obesity, cancer, knee problems, flat feet, shin splints, and lower-back pain. And that suddenly touched a raw nerve with me, because I realized that by studying the evolution of the human body we can address problems that have seemed intractable.

For example, many people are afraid of running because between 30 and 70 percent (depending on how you measure it) of runners get injured every year. The most common injury in the world is lower-back pain. Something like 70 or 80 percent of people today get lower-back pain and almost all of it is what's called nonspecific—that is, we don't know what caused it.

We often say that the reason people get lower-back pain is because we became bipeds and being a biped is a stupid way to use your back. But that doesn't make sense, because if back pain is so difficult, such a challenge, natural selection surely would have acted to lessen its prevalence and severity. In fact, if you start asking people who work with hunter-gatherers, most people say, "Yes, actually, come to think of it, I don't recall anybody saying they had back pain." I've never seen anybody have back pain in the hunter-gatherer context.

Our lives are filled with problems like insomnia and constipation which are extremely recent. They're novel, and they're caused by the way we misuse our bodies. The research I'm doing now is about how we can use our bodies better, particularly in terms of the musculoskeletal system, to avoid injury, pain, and the debilitating disabilities that prevent people from exercising and staying fit.

I would argue that many of the ways in which we get sick today have a corporate, almost capitalist origin. If you think about the obesity crisis: People are getting sick today because we've created industrial food that makes sugar unbelievably cheap, that makes low-quality fats unbelievably cheap. We evolved profound, deep, serious cravings for fat and sugar because those used to be limited and important resources in our evolutionary history.

It's also true for something as simple as shoes, or sofas, or elevators. We like to take it easy. Typical hunter-gatherers were always at the edge of energy balance. They could barely get enough food to satisfy their needs and their family's needs, so it makes sense, when you're a hunter-gatherer, to take it easy when you don't have to work. We're programmed to take the elevator or the escalator whenever we see it. In fact, there are experiments that show if you have a stairway next to an escalator, only about 3 percent of people will take the stairway instead of the escalator. People love to take the escalator. It's just hard-wired into our brains, I believe. That makes sense in a context in which you had to work hard and struggle to maintain energy balance.

Shoes are another interesting example. We love comfort. We have this idea that things that are comfortable must be good for us. So people buy shoes that are comfortable. Well, since when was there a relationship between comfort and health? I would argue that a lot of shoes actually cause people to become injured because they're comfortable. An arch support in a shoe is comfortable because that arch support means that the muscles in your foot no longer have to work anymore to support your arch. It's like taking the elevator all day long. Those muscles then atrophy, or they never develop properly, if you give kids arch supports. Their arches don't develop properly, or they collapse pretty quickly. Twenty-five percent of Americans have fallen arches, which is an amazing statistic.

In the Kenyan villages where I work, where people don't wear shoes, I have yet to find a single person with a fallen arch. They just don't exist. Maybe we'll find one eventually, like a black swan. But they're obviously extremely rare, those kinds of foot problems. They may have all kinds of crud in their feet and they have other shoe problems and foot problems, but collapsed arches don't seem to exist in barefoot populations.

We have been marketed and sold all kinds of products that we'll willingly buy because they're comfortable. Air-conditioning makes us comfortable, but is it good for us? Probably not. The list goes on. Comfortable chairs, for example. Just think about how bad chairs are for us today. Paper after paper, study after study, has shown that chairs give us back problems, because they shorten our hip flexors and give us weak backs. They make us sedentary. We take years off our lives, probably, by sitting in chairs, but we like them because they're comfortable. You go to an African village, you find me a chair with a back. That's a rare thing out there. We love comfort, and people make a lot of money selling us comfort, but I would challenge the notion that comfort is usually good for us.

One important question is, Can we test the idea that a more evolutionarily informed way of using our bodies is relevant to helping people. One idea is that a more barefoot style of running is less injurious. If you study barefoot runners, not only do they tend to land on their forefoot but they tend to take shorter strides, they have better posture, and there are a number of other things that are different from typical shod runners.

As a way of testing that, we measured the Harvard track team. We looked at all the people who run middle- and long-distance on the Harvard track team. Some of these kids are running 40, 50, some of them are even running 100 miles a week, really fast. It turns out about a third of them are forefoot strikers. They run with more or less what I would say is a good barefoot style. It turns out also that the Harvard track team has a great doctor and a trainer who have recorded every single injury. When they get scratched, or even just get a shin splint or a muscle tic—anything that happens to these kids gets written down and is diagnosed by a professional. The track team coach also requires them to log their mileage every single day—how far they ran and at what speed.

For some of these kids we had four years of data. So we took a high-speed camera, measured how they ran, and then compared the ones who ran with a forefoot kind of barefoot style versus those who landed on their heel. We found that we could assess how severe each injury was because we could measure how much effect each injury had on how many days they had to take off, and we came up with a quantified injury-severity score. When we quantified injury in a sensible way like that, the runners who were forefoot strikers had 2.6-fold lower injury rates than the heel strikers. It's a huge difference. As far as I know, it's the biggest effect ever shown on running injuries, and it's a perfect example of how taking an evolutionary approach to the body gives us insights about how to better use our bodies.

12
Mapping The Neanderthal Genome

Svante Pääbo

[July 4, 2009]

Svante Pääbo is the director of the Max Planck Institute for Evolutionary Anthropology.

We are now in the process of analyzing the Neanderthal genome, putting together all the little DNA pieces we have extracted from the sequence of this fossil and starting to compare it to humans and chimps. One question we are struggling with and thinking about is, What is the relationship of the Neanderthals to us?

We try to address that in different ways. One thing we're beginning to see is that we are extremely closely related to the Neanderthals. In a way, they're like a human ancestor 300,000 years ago. Which leads you to ask, "What if they had survived a little longer and were with us today?" After all, they disappeared only around 30,000 years—or 2,000 generations—ago. Had they survived, where would they be today? Would they be in a zoo? Or would they live in suburbia? These are questions I like to think about. It's the questions, and not the answers, that are interesting, because these questions have no answers. We will never know. But they're interesting questions because they reflect how we think about differences between us and our ancient ancestors.

If the Neanderthals were here today, they would certainly be different from us. Would we experience racism against Neanderthals, worse than the racism we experience today amongst ourselves? What if they were only a bit different, but similar in many ways—in terms of language, technology, social groups? Would there still be this enormous divide we see today between humans and nonhumans? Between animals and ourselves? I don't know how it would play out if Neanderthals were alive today. It's probably true that the tendencies that create racism in us would be at play in a world with Neanderthals—equally or even more. It's probably also true that the dichotomy between humans and all other mammals would not be as severe as it is today.

But it's interesting to speculate about the mutual genetic influence of humans and Neanderthals, in both directions. There's no unequivocal evidence that Neanderthals contributed genes to people living today, but that's not to say that such a contribution doesn't exist. The only thing we know is that they did not contribute mitochondrial DNA—that's the one thing we can say.

Just a few weeks ago, we discovered a large part of the Neanderthal genome—66 to 70 percent of it—so we can now address the question of their genetic contribution to modern humans much more rigorously. And we can ask about both directions of genetic influence, because, of course, mixture goes two ways. When we were able to study only modern humans—humans that live today—we couldn't find evidence of a Neanderthal contribution. But now that we have the Neanderthals, we can ask about the other direction. Is there any evidence that early human ancestors interbred with Neanderthals and contributed genes to them? That's something we are struggling with at the moment. It's a difficult analysis, because if tiny parts of a [Neanderthal] genome were contributed by humans, you have to assume that they weren't due to some error in our analysis, or contamination by modern-human DNA, or any bias in the algorithms we use.

In the public media, we're pretty much depicted as saying there was absolutely no mixture. It's hard to convey subtle messages to the public. If you read our papers, we say very carefully that there is absolute proof they [Neanderthals] didn't contribute *mitochondrial* DNA. That doesn't mean they couldn't have contributed other parts of their genome.

It's still clear that any genetic contribution from the Neanderthals to modern humans has to be pretty small. There's a lot of evidence, in the sense that Africa is more genetically diverse; there's more variation in Africa than for all humans outside Africa, although there are fewer people in Africa—only 800 million or so.

Everything we find outside Africa has close relatives inside Africa, in terms of genetic variants. I like to say that if we look at our genome, our DNA, we are all Africans today. Either we live in rather recent exile, maybe since 50,000 years or so, or we live inside Africa. If Neanderthals had contributed a lot to modern Europeans, Europeans would have genetic variants not found in Africa today, or that would differ a lot from Asians, for example—which we don't find.

As a paleontological outsider, I'm often surprised at how much paleontologists fight. Why is that the case? Why do we have less vicious fights in, say, molecular biology? I suppose the reason is that paleontology is a data-poor science. There are probably more paleontologists than there are important fossils in the world. To make a name for yourself, you must find a new interpretation for those extant fossils; this always goes against some earlier person's interpretation. There are other areas of science where we agree to disagree, but at least we often generally agree on what data we need to collect to resolve the issue. And no one wants to come out too strongly on one side or the other, because the data could in a year or two prove you wrong. But in paleontology, you can't decide what you'll find. You cannot, in most cases, go out and test your hypothesis in a direct way. It's almost like social anthropology or politics—you can win only by yelling louder than the other person or sounding more convincing.

Of course, regarding the question of modern human origins, the fight is between the multiregionalist versus the out-of-Africa hypothesis. Genetic data from studies of mitochondrial DNA in the 1980s and then data from other parts of our genome strongly favored the out-of-Africa hypothesis, which [Chris] Stringer was long one of the main proponents of in paleontology. It's clear from the genetic evidence that the big picture is the out-of-Africa one. That's not to say there couldn't have been a small contribution from earlier archaic forms such as Neanderthals in Europe to present-day Europeans, or *Homo erectus* forms in Asia to Asians, but it has to be very small. The other camp, which follows [Milford] Wolpoff, is in the minority today.

But the Neanderthal genome will be a chance to address that question. And it depends on what you're interested in. As a geneticist, I'm not so interested in who had sex with whom 30,000 years ago. As a geneticist, the question is, Did Neanderthals contribute significantly to our gene pool today? Did they have an effect on the variants we carry? Any effects have got to be small.

But for understanding what happened when modern humans met Neanderthals, this question of how we interacted is important. It would be interesting, for example, if we were to find that there was a gene flow, but in the other direction—mainly from modern humans into Neanderthals. When two groups meet and there's social inequality, there is almost always mixing but it's generally directional: Most often the males from the dominant group have offspring with females from the non-dominant group, and the offspring stay with the non-dominant group. If something like that happened when we met the Neanderthals, there would perhaps have been an inflow from modern humans into Neanderthals, and we would not see it in the mitochondrial DNA of modern humans.

But when we speculate about the Neanderthals, I often like to say that it's more about our worldview than anything that happened back then. If you're a racist, you could play it either way: You could say that if Neanderthals contributed to modern Europeans, there must be old variants, adapted to living in Europe, which had been there for hundreds of thousands of years. There was this group adapted to living in Europe, say, which then started to move around the world. You can start telling stories like that. But you can equally say that the people who left Africa were the more innovative, advanced people who exploited new territories—that this was somehow in the genetic subset of what existed in Africa. You can spin it either way you like. I don't think there's any scientific knowledge or insight that will convince people to change their ingrained ideas about this. One is often asked, "Why do you think Neanderthals disappeared? Did we kill them all?" If you like to see modern humans as violent—just look at how we behave today—you'd say this was obviously our first big genocide.

On the other hand, what if I say that the earliest modern-human fossils in the Middle East are 93,000 years old? The latest Neanderthals in the Middle East are 60,000 years old. That means we have 30,000 years of peaceful coexistence in the Middle East. (If we could have that today, it would be wonderful!) You can say, "Well, that's probably because modern humans came, and the Neanderthals disappeared, and modern humans disappeared again, so they never interacted with each other." But who knows? The speculation just reflects how we think of ourselves.

The first attempts to extract DNA from old remains go back to the first half of the 1980s, when I started working with ancient Egyptian mummies. But it was the invention of polymerase chain reaction, PCR, by Kary Mullis that made it possible to target a piece of DNA you were interested in from a fossil and reproduce your results and show that you could do it again and again—that it was reliable. Others could repeat it. That was in the late 1980s and the beginning of the 1990s.

And this was applied to the Neanderthals for the first time in 1997, which was when we started to have an impact on paleontology and paleoanthropology. But over the past two years a new technology has arisen: high-throughput DNA sequencing. There are machines now where you can take the DNA you've extracted from a fossil and sequence random pieces of it so efficiently that you can, without targeting anything special, see everything that's in there and then look at what, on this DNA molecule, looks similar to that of humans or chimps. And it's generally just a few percent—2, 3, 4 percent—but there's such a tremendous throughput in these technologies that you can afford to do that. You can throw away 95 to 98 percent of your data and just look at the rest. That has changed the ballgame, so that we can now begin to look at the total genome of extinct forms, such as the Neanderthals or the mammoths or other extinct animals.

As a kid, I wanted to be an archaeologist and Egyptologist and I wanted to excavate in Egypt and things like that. As is often the case, I had a far too romantic idea about these things. When I got to university, I started studying Egyptology and found that it was not at all like Indiana Jones, as I had imagined. At least as it was taught in Sweden, it was very much linguistics—thinking about ancient Egyptian verb forms and things like that. I got disenchanted with it and didn't know what to do. I was also influenced by my father, so I decided to study medicine.

But that was when DNA technologies and cloning were coming of age. I knew there were thousands of mummies in collections in Egyptological museums and that hundreds of new mummies were found every year in Egypt. No one seemed to have applied the new technologies to them, which was rather the obvious thing to do—that is, take a sample of an ancient Egyptian mummy, extract the DNA and clone it in bacteria to study it.

I started doing that as a hobby. I was a bit scared of my thesis adviser, who was a rather domineering person, so I did it secretly at night and on the weekends. It was a successful endeavor. We showed that you could stain DNA in the cell nuclei in the tissue samples from a few Egyptian mummies. You could also extract the DNA and show that it was human. But you were then hampered by the impossibility of retrieving any specific things, so small were the traces of DNA. That changed two years later, when PCR came about. Then I ended up going to Berkeley, to a lab that also had interest in this, and that's where I developed it.

When I started out, 1984-85, intent on studying the genomes of ancient civilizations, I was driven by delusions of grandeur. I thought I could easily study the ancient genomes. I dreamt of addressing questions in Egyptology: For example, how did the historico-political events we read about affect the population? When Alexander the Great came to Egypt, what was the influence on the population? Was it just a political change? The Arab Conquest: Did that mean a large part of the population was replaced? Or was it mainly a cultural change? There was no way we could answer these questions from historical records, and my dream was to address questions like this. After some initial success, I realized the limitations on what I wanted to do.

There was then a much longer period in which I concentrated on extinct animals. We did the first DNA sequences from a mammoth, from marsupial wolves, from the moas in New Zealand. We didn't have to struggle with the contamination issue there; modern-human DNA can easily be distinguished from those things.

With a few new technological possibilities, I can begin realizing the dreams I had in the 1980s. One big dream is to address questions specific to humans relative to other life-forms. Such as language. So it was exciting a couple of years ago when the FOXP2 gene was identified. A mutation in that gene in humans resulted in a specific speech problem, seeming to do with articulation. The primary problem concerns muscle control in the oral thorax—a millisecond of control you need over what your vocal cords, your tongue, your lips do to produce articulate speech.

We studied the evolution of that gene. The protein that's made from it is a protein whose function is to turn on and off the activity of other genes in the body. And the protein carries two amino-acid substitutions unique to humans. They're seen in no other primates; this happened in the human lineage.

There are also patterns of variation in the FOXP2 gene among humans today suggesting that there had been positive selection acting on it and that one variant spread rapidly to all humans on the planet. It was tantalizing to speculate that those variants were those amino-acid changes and that they had an effect on our ability to produce articulate speech.

There are two issues we are exploring since we addressed the variation question and showed a singular positive selection. One is to look at the Neanderthals and determine whether they had these amino-acid changes or not. And it turns out that they did share these amino-acid changes with us—which I found surprising. To the extent that the changes have something to do with articulate speech, we share that with the Neanderthals. This is, of course, one gene out of many others still unknown having to do with language and speech. There could have been some difference [in speech ability], but from the little we know, there's no reason to assume a difference.

The other question we are addressing is whether these amino-acid differences are of importance. What we have done is construct a laboratory mouse that makes not the mouse version of this protein but the human version, from its endogenous FOXP2 gene—mice, like every other vertebrate, have a FOXP2 gene. We've analyzed that mouse extensively over the last two years and run all kinds of tests. We and our collaborators have looked at over 300 different traits in this mouse, always comparing the knock-in mice that are changed—the humanized mice—to littermates born of the same mother that are wild-type producing the mouse protein. We compare them directly to each other—so they will have had the same birth experience, the same environment. There are only two of over 300 traits we've looked at where the humanized mice differ significantly from the wild-type mice.

One thing I don't understand is that the humanized mice are slightly more cautious in a new environment than the wild mice. If, for example, they come to an open area, where a mouse feels more exposed and vulnerable, the humanized mice stay along the walls for the first few minutes, whereas the wild-type mice are bolder and enter the open area. But that's a difference that lasts for the first four or five minutes, and then there's no difference. I don't know what to think about that.

The other mystery is tantalizing and shocked me: the mice vocalize differently. We measure that by taking the pups away from their mother when they're two weeks old; they peep so that the mother comes and brings them back into the nest. We can record that in the ultrasound area. We work with people who are experts in analyzing sonograms of vocalizations, and there is a clear— subtle, but clear—difference in vocalization. This supports my belief that these changes have something to do with muscle control in the oropharynx or so, probably something to do with articulation in humans. But that's an earlier change in the human lineage than the divergence to the Neanderthals; it's something we share with the Neanderthals.

The major thing we realized early on when we started applying PCR in ancient DNA is that contamination is a serious issue—particularly contamination by modern-human DNA, because it's all around us. The dust in this room is to a large extent skin fragments from our bodies that contain DNA and can land in experiments or be in the chemicals we use. That's why we decided to stay away from studying modern-human DNA early on, such as from ancient Egyptian human mummies, for example—because it was almost impossible to show that you had the right thing. With Neanderthals, it's a bit different, because with regard to the mitochondrial DNA we have clear differences.

When we applied high-throughput sequencing technology to the Neanderthal remains, we tried different technologies. We made two extracts from the best Neanderthal we had, from Croatia, and we sent one extract to Eddy Rubin's group at Berkeley, who cloned it in bacteria. They used technology that had been around from the 1980s but was now more efficient, sequenced clones, and retrieved 60,000 to 70,000 base pairs from the Neanderthal. We sent another extract from the very same bone to 454 Life Sciences, in Branford, Connecticut, to Jonathan Rothberg, who founded the company and bought the rights to pyrosequencing from Sweden. Then we applied their technology to it, and that turned out to be much more efficient. They sequenced almost a million base pairs—around 750,000. We analyzed these different data sets and there were two papers published in the same week; one in *Science*, which Eddy Rubin produced, and one in *Nature*, produced with 454. [Noonan, J. P., et al., "Sequencing and analysis of Neanderthal genomic DNA," *Science*, 314, 1113-1118 (2006); Green, R. E., et al., "Analysis of one million base pairs of Neanderthal DNA." *Nature*, 444, 330-336 (2006).]

There was some tension with Eddy Rubin's group, because they were intent on continuing with the cloning approach in bacteria. But it seemed obvious to me that it wasn't efficient enough; they had produced 10 times less data from what we had sent them than 454 produced. . . . And there was contamination in the 454 data set; we still don't know where it entered, since both extracts came from our clean room, prepared from the same bone. The most likely thing was that this was introduced at 454 at the same time when they were sequencing Jim Watson's genome. I wouldn't be totally surprised if there was some slight mixture of Jim Watson with the Neanderthal sequence there.

In 2008 we published a paper in *Cell* [Green, R. E., et al., "A complete Neandertal mitochondrial genome sequence determined by high-throughput sequencing," *Cell*, 134(3), 416-426 (2008)] with the complete Neanderthal mitochondrial genome, where we then discussed this point. . . . We stated that we would tag the sequences in the clean room so that when they left the clean room for sequencing, every sequence would start with this little tag. That way we would know the sequence came from our clean room. Because we do see that when we use small amounts of DNA, there is some carry-over in the sequencing machines.

The next big thing as far as announcements coming out of

my lab is the scientific publication of the first draft version of the Neanderthal genome. The only finished genomes among mammals are the human and the mouse. Everything else is drafts. The Neanderthal will be a very, very rough draft—it's just 1.5-fold, meaning that statistically every nucleotide in the genome has a 1.5-fold chance to have been hit. But that means there are a lot of parts of the genome that are not hit at all—something like 30 percent we have not seen. But we can get a first overview of the genome. We can make windows of, say, 100,000 base pairs, and out of those we will get, on the average, 60,000, 70,000, and go over the genome. What's involved in getting to 100 percent is sequencing much deeper in more samples. We will be doing this in the next few years.

People ask me why I ended up in Germany. Like so much that happens in life, it was by chance. When I was at Berkeley, I wanted to go back to Europe, and I had a girlfriend who was a graduate student in Munich, so I visited her. Her professor, a very good geneticist, asked me to give a seminar and said there would be a position open in a year if I wanted to apply. Things move slowly in university politics; by the time I got an offer, I didn't have that girlfriend anymore. But it ended up being a good offer compared to what was available to me in Cambridge or Sweden. I thought, "Well, it's not that I ever dreamt about living in Germany for the lifestyle, but I can go there for a few years, do good work, and then move on." And I learned that that's a very good attitude—i.e., to encounter places you don't expect to be Paradise. My experience of Germany is that it's a lot nicer than my preconceptions. I had a good time there, and then six or seven years later came the opportunity to be part of the process of opening a new Max Planck institute in Eastern Germany, where there was a political decision to build research institutes to the same density as in West Germany after reunification.

The Germans were willing to ask in what areas of research Germany was particularly weak. And obviously one such area was genetics, and anthropology in particular, and for good reason. The Max Planck Society, the predecessor of the Kaiser Wilhelm Society, had an Institute of Anthropology in Berlin, where people like Josef Mengele, who was heavily involved in the Holocaust, had appointments. So since May 1945, the Max Planck Society of course stayed away totally from anthropology. And that was also reflected in the university landscape; anthropology, after the war, had no prestige at all as an academic discipline, and the quality was correspondingly low.

Creationism is a much smaller issue in Germany than in America. Most of the e-mails and letters I get in that vein come from America, much more than from Germany. The few critical comments we have received have come from that quarter; they object to the idea that we share an ancestor with Neanderthals, or with chimpanzees. But I have been able to work with religious fundamentalists—mostly a graduate student in my lab from the Middle East. In the end, I can't argue with someone who says, "God is almighty, and anything I perceive or think can be put in my head by God, because God can make me think there is evidence for evolution." I can't argue against that. I couldn't understand why a God would deceive me in such a way, but if that's the case, then we study God's deception, right? We agreed on that with this graduate student, and then we could work in that framework.

It's much more the case in the U.S., that when you give talks or attend evolutionary conferences, there can be organized opposition to what's happening. There's no place for religion in scientific conversation. Science has to stay science, and religion religion. Obviously religion has an important role in the lives of probably 80 percent of the people on the planet today, and it's an important one. Science need not go out and fight it, but there's no role for religion in science. These are two different realms. I must admit that if I'm confronted with existential questions or life-and-death things, I tend to become irrational and think in magical ways. Often when we are confronted with things that threaten the foundations of our life, we tend to go for religious or magical or nonscientific ways of dealing with it. That's part of being human.

13
On Biocomputation

J. Craig Venter, Ray Kurzweil, Rodney Brooks

This live *Edge* event was presented on February 23rd, 2005, hosted by the TED Conference (Technology, Entertainment, Design) in Monterey, California.

Ray Kurzweil, author of *The Singularity is Near*, is one of the world's leading inventors and futurists.

Rodney Brooks is Panasonic Professor of Robotics, Emeritus, at MIT.

J. CRAIG VENTER

With such a broad topic I had no idea where to start, so I decided to start with the whole planet's genome. As many of you know, the human genome was sequenced in the year 2000, and that gave us the complete repertoire of human genes. And slowly, as we're adding more and more gene space to that—and we've gone up 2 orders of magnitude just in the last year, from new genes in the environment—we're starting to look at the world in terms of gene space instead of genomes and species, and this gets us down to component analysis.

This is affecting three different major areas we're working on at the Venter Institute. Cancer is the one I'd like to address the most today, but I'll mention the others as well. Cancer is getting broken down into more and more separate diseases, as we're able to subdivide the diagnostics and the genes associated with them, but we're starting to take a different view, viewing cancer as an overall disease, and we're looking at gene space—where we think it can be targeted to deal with cancer as a whole. A lot of cancer's been looked at genetically, and while there are genetic predispositions to cancer that we all have, it's actually somatic mutations that we get from toxins in the environment—from radiation, etc.—that usually lead to cancer. The model is that as we accumulate mutations in more and more genes, we cross a threshold where all of a sudden we have unregulated cell growth.

It turns out that there's a set of about 518 genes, kinase receptors, that are responsible for controlling cell growth. What we found is, obviously we could identify all those in the genome now. We could look for mutations in individuals with cancer, whether they inherited those mutations, but now that we can do high-throughput resequencing of genes, we're now sequencing these genes looking for somatic mutations, things that have occurred after the genome was established. And in every type of cancer, we're finding mutations in these genes. Usually these mutations lead to unregulated cell growth. It turns on the kinase receptors and they run continuously.

There have been some remarkable breakthroughs in the last few years with just a couple of drugs that block these receptors and interfere with the growth of cancer. Herceptin, out of Genentech, is a therapeutic antibody that affects one of these receptors. Probably the most important breakthrough has been Gleevac, which works by blocking the rapid growth of white blood cells and has led to almost miracle cures of cancer. There's another drug that Novartis sells that initially in trials didn't look very good, until some of the receptors were sequenced in the cancer patients, and it turns out that 10 percent of patients they looked at had these mutations in this receptor, and the drug worked on virtually 100 percent of those individuals.

By understanding this gene space, understanding mutations that we've collected during our lifetimes, we may be able to have a set of molecules that work universally against if not all, certainly most cancers. And there are a few other groups of molecules that seem to fit into these categories. So we have a major new program; we recruited Bob Strausberg of the National Cancer Institute, in collaboration with the Ludwig Institute for Cancer Research, and with Bert Vogelstein at Johns Hopkins, trying to look at as many different cancers as we can, looking at somatic mutations in these. Once we collect a big enough set of these, it's easy just to design a gene chip to turn this into a few-dollar assay for people instead of expensive gene sequencing. So we're trying to use the bioinformatics to predict other gene sets that look like they're in the same category, to see if we could basically have on the shelf a repertoire of small molecules and antibodies that would work against most types of cancer. The excitement on that front is pretty stunning. Most of these things represent a change in philosophy.

We're taking the same approach to antivirals and antibiotics (with all the worry about bioterrorism, we have very few antivirals) and beginning to look at common mechanisms and infection. We have choke points that we think can block these infections regardless of whether it's an Ebola virus or a SARS virus. There are a lot of groups now effectively working on this. Understanding gene space and what it starts to mean in a post-genomic era is giving us a lot of new insight.

We're adding, as I said, exponentially to the number of genes by just doing shotgun sequencing of the environment. There were 188,000 well-characterized genes in the protein databases. We're up to close to 8 million new ones just from doing random shotgun sequencing from the oceans. And we took all this combined data set and then we tried to see how many different gene families we have on the planet—trying to get down to our basic component sets. The number right now is somewhere between 40,000 and 50,000 unique gene families, covering all the species we know about. But every time we take a new sample from the environment and sequence it, we keep adding to those gene families in a linear fashion, seeing that we know very little of the overall biology of our planet. But each of these new genes—and we have literally several million genes of unknown function—some of these, we have families over 1,000 related proteins or genes in the same family. They're obviously important to biology, they're important to evolution, and beginning to sort out, using bioinformatic tools, what they may do, is giving us a lot of new tools in different areas.

The third and last area is synthetic biology. Trying to understand the basic components of a cell, we've tried knocking out genes to see what gene cells could live without, but we get different answers every time the experiment's done, depending on how it's done—whether it's a batch growth or you require cloning out of the cells, different growth requirements. We decided some time ago that the only way to approach this was to build an artificial chromosome and be able to do evolution in the laboratory the way it happens in the environment.

We're building 100 cassettes of 5 or more genes, where we can substitute these cassettes, build an artificial chromosome, and try and create artificial species with these unique sets. But now, with 8 million genes, and as this work continues, it's conceivable within a year or two we'll have databases of 30, 40, 100 million genes. Biology is starting to approach the threshold that the electronics industry passed, where all of a sudden people had all the components and could start building virtually anything they wanted using these different components. We have a problem; we don't understand all the biology at first-principle levels yet, but we're getting the tools, we're getting the components where we can artificially build these. And we think we can, in the computer, design a species, design what biological functions we want it to have, and add this to the existing skeleton.

Understanding the gene components, working forward from those, we're applying this to energy production. We've tried to change photosynthesis by taking oxygen-insensitive hydrogenases, and we're converting all the electrons direct from sunlight into hydrogen production. We're doing this with a molecular switch, so you can throw the switch and hydrogen bubbles off and you turn the switch off chemically and it stops the production. We're also trying to come up with new ways for fermentation from wood. So we're approaching things on a broad level, looking at genes as the fundamental components of biology and the future of industry.

On a more specific level, bird flu is currently in the news again. It's a good thing to avoid. In fact we're working with the prime minister of Thailand, we're working with Hong Kong, trying to use these same sequencing tools to track these new infections. The trouble is, the flu virus—when you get, for example, two or more isolates in a pig—can recombine to form an essentially infinite number of new viral particles. And those transfer into humans at a frequency we don't like. And with the constant development of new viruses in birds which transfer these to animals, it's a matter of tracking these before we have a new pandemic.

People saw what happened with SARS: With air travel, you get an outbreak in China, the next thing it's happening in Toronto. By tracking the sequence space in birds around the world, we're trying to develop programs where we can catch things at an early enough stage with early detection programs, so that sequencing the gene space, even though we don't understand why the 1914 pandemic virus was so virulent — by understanding the components that are out there and tracking how they recombined — we hope to be able to avoid a new pandemic.

RAY KURZWEIL

Let me try to build on what Craig has said. We just heard about some exciting applications that are in the early stage, moving on from the general project—where we essentially collected the machine language of biology and we're now trying to disassemble and reverse-engineer it. And I come to this from a couple of perspectives. I've actually had these two disparate interests in my life. One has been computer science. Both Rodney and I have worked in the AI [artificial intelligence] field. And then I've had an interest in health. It started with the premature death of my father when I was twenty-two. I was diagnosed with diabetes in my mid-thirties. Conventional treatment made that worse. I came up with my own program; I've had no indication of diabetes since. I wrote a health book about that in 1993.

Thus, I've had an interest in health. My most recent book is about health, where we talk about three bridges to being able to radically extend our longevity. Bridge #1 is what we can do today—we can actually do a lot more than people realize—in slowing down degenerative disease processes, and to some extent aging, which can keep even baby boomers, like my co-author and myself, and this panel, in good shape, until we have the full flowering of this biotechnology revolution, which is the intersection of my main interest, Rodney's main interest, which is information technology, and biology.

We call that the second bridge. That can then extend our longevity until the third bridge—the full flowering of the nanotechnology revolution, where we can go beyond the limitations of biology, because even though biology is remarkable in many ways, remarkably intricate, it's also profoundly limited. Our interneuronal connections in our brain, for example, process information at chemical signaling speeds of a few 100 feet per second, compared to a billion feet per second for electronics. Electronics is a million times faster.

There's a robotic design for red blood cells by Rob Freitas which he calls respirocytes. A conservative analysis of that indicates that if you replace 10 percent of your red blood cells with these devices, you could do an Olympic sprint for 15 minutes without taking a breath, or sit at the bottom of your pool for four hours. Our biological systems are clever, but they're suboptimal. I've watched my own white blood cell under a microscope, and it had intelligence. It was able to notice a pathogen—I was watching this on a slide. It cleverly blocked its exit, cornered it, surrounded it, and destroyed it, but didn't do it that quickly; it took an hour and a half—a very boring thing to watch. There are proposals for circa 2020s technology of little robots the size of blood cells that could do that hundreds of times faster. These may sound like futuristic scenarios, but I would point out that there are four major conferences already on bioMEMS—little devices that are blood-cell size and already performing therapeutic and diagnostic functions in animals. For example, one scientist has cured type-1 diabetes in rats with a nano-engineered device with seven nanometer pores.

In terms of my health interests, this biotechnology revolution is the second bridge. And we're in the early stages of it now, but

10-15 years from now many of these technologies Craig mentioned—a few of the many examples that are now in process—will be mature.

My other interest is information technology. I am an inventor, and I realized that for inventions to succeed, they have to make sense for the world when you finish the project—and most inventions fail because the enabling technologies are not in place. Thus I became an ardent student of technology trends and began to develop mathematical models of how technology evolves. The key message here is that information technology in particular progresses exponentially.

Craig certainly has experienced this; the genome project was controversial when it was first announced. Critics said, "How are you going to get this project done in 15 years? At the rate at which we can sequence the genome and with the tools we have, it's going to take far longer." Two-thirds of the way through the project, the skeptics were still going strong, because not that much of the project had been finished. I'll show you charts of the exponential reduction in the cost in sequencing DNA over that period, and the exponential growth in the amount of DNA that's being sequenced. It took us 15 years to sequence HIV; we sequenced SARS in 31 days. There's been smooth exponential growth of that process. People are familiar with Moore's Law and some say it's a self-fulfilling prophecy, but in fact it's a fundamental attribute of any information technology. We create more powerful tools, those tools then are used for the next stage. Often scientists don't take into consideration the fact that they won't have to solve a problem for the next ten years with the same set of tools. The tools are continually getting more powerful.

The other major observation is that information technology is increasingly encompassing everything of value, from biology to music to understanding how the brain works. We have information in our brains, and even though some of it's analog rather than all digital, we can model it mathematically—and if we can model it mathematically, we can simulate it. We're further along in reverse-engineering our brains than people realize. The theme of this panel is the intersection of information technology and biology, and that's a new phenomenon.

Craig was the leader of the private effort to sequence the genome that really launched this biotechnology revolution. We are in the early stages of understanding how biology works. Most of it we don't understand yet, but what we understand already is very powerful, and we're getting the tools to manipulate these information processes. Almost all drugs on the market today were created through what's called drug discovery, which is where pharmaceutical companies methodically go through tens of thousands or hundreds of thousands of substances and find something that seems to have some benefit. "Oh, here's something that lowers blood pressure!" We have no idea how or why it works, but it does lower blood pressure—and then we discover it has significant side effects. Most drugs were done that way. Kind of like the way primitive man or woman would find tools. "Oh, here, this stone would make a good hammer!" We didn't have the means of shaping tools. We're now gaining those means, because we're understanding how these diseases progress. In the case of heart disease, we already have a mature understanding of the sequence of specific biochemical steps and information steps that lead to that disease. Then we can design drugs that intervene precisely at certain steps in that process.

There are literally thousands of developments in the pipeline, where we can rationally intervene, with increasingly powerful tools, to change the progression of information processes that lead to disease and aging. The tools include enzyme inhibitors: If we find that there's an enzyme critical to a process, we can block it. We also have a new tool called RNA interference, where we send little fragments of RNA into the cell—they don't have to go into the nucleus, which is hard to do—and they basically latch onto the message-RNA representing a gene and destroy it, and it inhibits gene expression much better than the older anti-sense technology. That's very powerful, because most diseases use gene expression someplace in their life cycle, so if we can inhibit a gene we can circumvent undesirable processes.

I'll give you just one example: the fat insulin receptor gene basically says hold on to every calorie because the next hunting season may not work out so well—which points out that our genes evolved tens of thousands of years ago, when circumstances were very different. For one thing, it wasn't in the interest of the species for people to live much past childbearing age, and people were grandmothers by the age of thirty. Life expectancy was thirty-seven in 1800. So we've already begun to intervene, but we now have much more powerful tools.

When scientists at Joslin Diabetes Center inhibited the fat insulin receptor gene in mice, these mice ate ravenously and remained slim, got the health benefits of being slim—they didn't get diabetes, they didn't get heart disease, they lived 20 percent longer. They got the benefits of caloric restriction without the restriction. Some pharmaceutical companies have noticed that that might be an interesting drug to bring to the human market. There are some challenges, because you don't want to inhibit the fat insulin receptor gene in muscle tissue, only in the fat cells; there are some strategies for doing that. But it's an example of the power of being able to inhibit genes, which is another tool to basically re-engineer these information processes. The way we easily do with a Roomba vacuum cleaner, we can just change the software, but we're actually gaining the means to do that now in our biology.

More powerfully, we'll be able to add new genes. Until recently, gene therapy has had challenges in terms of getting the genetic material into the nucleus and also getting it into the right place. There are some interesting new strategies. One is to collect adult stem cells from the blood, then in the petri dish insert the new genetic material; discard the ones that don't get inserted in the right place, and when you get one that looks good, you replicate it and then reinsert it into the bloodstream of the patient. A project by United Therapeutics succeeded in curing pulmonary hypertension in animals, which is a fatal disease. It's now going into human trials. There are a number of other promising new methodologies for gene therapy.

We ultimately will have not just designer babies but designer baby-boomers. And there are many other tools to intervene in these information processes and reprogram them. There are no inexorable limits to biology. People talk about the telomeres—and say this means you can't live beyond a hundred and twenty. But all these things can be overcome through engineering. Just in the last few years, we've discovered that there's this one enzyme, telomerase, that controls the telomeres. These are complex projects. Somebody here is bound to say that we know very little about our biology. That's true. We had sequenced very little of the genome early in Craig Venter's project. But the progress will be exponential, and the toolswill be exponentially more powerful.

What we know already is providing a great deal of promise that we can overcome these major killers—like cancer, heart disease, type-2 diabetes and stroke. We're also beginning to understand the processes underlying aging. It's not just one process; there are many different things going on. But we can intervene to some extent already, and that ability will grow exponentially in the years ahead.

RODNEY BROOKS

I'm going to start off with some conventional stuff. The things Craig has done, and others, on sequencing genomes have really relied on algorithms that have been developed in the information sciences. And the work in genomics, proteomics, et cetera, that's going on now uses a lot of machine-learning techniques—statistical machine-learning techniques, developed largely out of work of theoretical computer scientists and then applied by clever people like Craig into biology.

There's a crossover between information science enabling the sorts of things Craig and others do. Interestingly, to me as a lab director, across a broad range of computer science, the impact of theoretical computer science is profound, but it's the hardest thing to get funded from external agencies. Because people who work in networking, or compilers, or chip design, or whatever, they'll write a proposal that says, "In the first three months we're going to do this task, in the next three months we're going to do that task," et cetera, and the funding agencies like that, because they see what they're going to get ahead of time. But the theoreticians are going to think about stuff. Maybe they'll prove some theorems and maybe they won't, but they don't say, "Oh, in the first three months we're going to prove these three theorems, then in the next few months we'll prove these others"—they would have already done the work. It's hard to get funding for theoretical computer science, because it doesn't fit that model of turning a crank and getting things out. But it has enormous impacts on biology.

What's happening now, though—and Craig mentioned some of this with synthetic biology—is that we're starting to move from just analysis of systems into engineering systems. I want to say a few words about engineering in general, and then about what's happening in biological engineering and how it will change completely from what people are thinking about now.

First on engineering. Engineering today is really applied computer science, in my view. Maybe that's a little biased, but essentially two things are going on in engineering. First, you analyze stuff—and these days that's all about application of computation, getting the right computation systems together to do the analysis—and second, engineering is also creativity, designing things, and these days it's designing the flow of information about how the pieces come together. From one point of view, all of engineering these days is about applied computer science. In that sense, as we go to biological engineering, it's just more applied computer science, applied to biology. Roughly around 1905 in electrical engineering is where we are today, in 2005, in bioengineering. Electrical engineering in 1905 looks very different from how electrical engineering looks today. In the next 100 years, what is now bioengineering is going to change dramatically. Ray, of course, is going to exponentially speed it up, but biology's actually more complex than physics, so there's going to be a little balance there and it may not go as fast as Ray assumes.

Let me go back to electrical engineering in 1905. Electrical engineering was just then split off from physics. It was applied physics, and at MIT it was actually in 1904 that the physics department got together and had a faculty meeting where they expelled "the electricals." That's the phrase they used in their minutes. There were these dirty "electricals" who were cluttering up their nice physics department. And that was the foundation of the electrical engineering department at MIT.

Today we're seeing the same sort of thing as bioengineering departments form. There was biology—applied biology, and now it's bioengineering, and that's happening in the engineering schools rather than in the science schools within the universities. But electrical engineering back in 1905 was a craft sort of thing; there wasn't the basic understanding that came 50 years later, when electrical engineering became science-based in the 1950s, and it changed the flavor of electrical engineering, and then in the last 50 years it's become information- and computer science-based.

Engineering transformed into applied computer science. Right now, bioengineering is starting to do a few interesting things, but it's only a shimmer of what will happen in the future. Craig mentioned hacking away at these mycoplasmas, hacking up genes, trying to get a minimal genome, and then getting pieces together and forming them into a synthetic biological creature. There are other groups that are going back even to pre-genetic approaches.

A group out of Los Alamos, funded through the European Union, is trying to build an artificial cell that doesn't necessarily rely on DNA; they're playing around with RNA and more primitive components. There's some sort of root engineering of trying to figure out how these biomolecules can fit together and do stuff they don't do in the wild. At our lab, in conjunction with a few other places, we've been working on something called biobricks, which are standard components, inspired. . . . If you go to the Web site parts.mit.edu, it's about biological parts. You see a 7400 series manual—people remember the 7400 series chips, which was what enabled the digital revolution, where you could put standard components together, so they've got a hacked-up version of the TTL handbook cover there—and it's biological parts—they're genes, a few hundred of them now; they've got part numbers, serial numbers; you click on the different sorts of part classes, you see a bunch of different instances of those parts, you see how they interact with each other.

We're running courses now where in January we have an independent activities period between the two semesters, and we get freshmen in, and they start clicking around with these parts, and they build a piece of genome which then gets spliced into an *E. coli* genome. *E. coli* is the chassis you build your stuff in, because it maintains itself and reproduces. And freshmen after two or three weeks are able to build engineered *E. coli*, which do things. Maybe they're an oscillator and they plug in a luminescence gene and they flash, very slowly. So you can build digital computational elements inside these *E. coli,* and it's not to replace computation in silicon, because the switching time on these things is about 10 minutes, using the digital abstractions—maybe using the digital abstractions isn't the right thing ultimately.

Or other sorts of projects that people have built—a sheet of *E. coli* sitting out there, and one of them will switch on and say, "Hey, guys," and start pushing out some lactone molecules, and the guys around that will sense that and then start clustering around the guy who said, "Come to me." It's the start of engineering living cells to do stuff they wouldn't do normally. It's going to change, ultimately, over the next 50 years, the basis of our industrial infrastructure.

If you go back 50 years, our industrial infrastructure was coal and steel. And in the last 50 years it's been transformed into an information industrial infrastructure. This engineering, at the molecular level, at the genetic level, of cells, is going to change the way we do production of a lot of stuff over the next 50 years. Right now you grow a tree, you cut it down, and build a table. Fifty years from now we should just grow the table. That's just a matter of time—and if we take Ray's point, it'll only be 15 years rather than 50—but I'm being conservative here. There's some stuff to work out, but it's just a matter of working through the details. We've seen, in broad strokes, how to do that.

That's biology and engineering at the molecular level. There's also biology at other levels, and we see a lot of that starting to happen at the neural level. So Mussa-Ivaldi and other people at Northwestern are using neural networks—biological neural networks—to control robots. There's a little wet stuff in the middle of a robot, controlling it. People at Brown and Duke are plugging into monkeys' heads, using the same machine-learning techniques used in genome analysis, to figure out what signals are going on in the monkeys' heads and letting the monkeys play video games just by thinking. Or control a robot just by thinking. And if you look in the latest *Wired* magazine, you'll see some human subjects, some quadriplegics, starting to be experimented with using this technology. So there's another case in which we're going from analysis of what happens inside to changing what happens inside. It's switching around the paradigm we use for biology from science to engineering.

But then there's going to be another level that happens over the next few tens of years. By playing with biology we're going to change the nature of engineering again; in the same way that over the last 100 years engineering has turned into an information science—an applied computer science—the way engineering is going to work is going to change again. The details of how it will change I can't say, because if I knew that, it would have already happened. It hasn't happened yet. But let me give you an example of the way biological systems are going to inspire engineering systems.

In my research group, we've been inspired by looking at polyclad flatworms. These are little ruffley sorts of things that—if you scuba dive and you've ever been on a coral reef, you've seen these little ruffley things move around. They're colorful and they've got ruffles around their edges. They're very simple animals. Their brain has about 2,000 neurons, and they can locomote and they can grab some food and, using their ruffles, push it into their mouth in the center. And I'm guessing—it never says in the papers, but in the 1950s a series of papers started, probably because of an accident that a grad student made.

They were seeing whether they could transplant brains between these polyclad flatworms. So they would cut the brain out of one, cut the brain out of the other, and then swap the brain, to see whether the function of these polyclad flatworms could be regained. When the brain was cut out, they were dumb polyclad flatworms. They couldn't right themselves, but they could locomote a little but not very well; if food came right near their mouth, or their feeding orifice, they'd grab it, but they couldn't pull the food in. Put another brain in, and a few days later they were almost back to normal.

Here's where the mistake happened: If you take the brain and you turn it 180 degrees around and put it in backwards, the flatworm doesn't do too well. It tends to walk backwards early on, but after a few days it adapts and it's reoriented and can do stuff. In fact, if you look at the geometry of these worms, there are two sets of nerve fibers on each side—there are four nerve fibers running along the length of the body, right through the brain. If you cut the brain out, it's got four stubs this way and four stubs that way, and if you turn it around 180 degrees, the stubs can line up and regrow and adapt. And if you flip it over, upside down, it works, too. If you flip it over and turn it around, it works—though some things don't quite work out. If you take the brain out and you turn it 90 degrees, it never works again. Because the stubs don't join up. If you take the brain out and cut a hole in the back of the worm, and put it down the bottom, it still works pretty well. This is just 2,000 neurons.

Imagine taking a Pentium processor out of an IBM PC and plugging it backwards into your Mac socket, and it working. That's not the sort of thing our engineering does today, but that is the sort of thing that biology does all over the place. We're going to see, by playing around and engineering these biological systems, a change in our understanding of complexity and our understanding of what computation is.

We have one model of computation right now, but I'm expecting—just as computation came along out of previous mathematics, without new physics or chemistry, just a rethinking of fairly conventional mathematics, and the notion of computation developed from around 1937, over the next 30 or 40 years—we're going to see some different sorts of understanding of complexity, and something which, by analogy—as computation was to previous discrete mathematics—we'll see this complexity understanding in relation to conventional information or computation understanding, and that then is going to change our engineering overall and the way we think about engineering over the next 50 to 100 years. It will all be the fault of understanding biology better.

KURZWEIL: Let me agree with a lot of what you said, Rodney. We have pretty similar views but we haven't, despite being on lots of panels, fully reconciled our models.

Let me address this issue of time frames, because it's not just a matter of hand-waving and noticing an acceleration. I have been modeling these trends for the past 25 years and have a track record of predictions based on these models. People say you can't predict the future, and that's true for certain types of predictions. Will Google stock be higher or lower than it is today three years from now, that's hard to predict. But if you ask me what will the cost of a MIPS of computing be in 2010, or how much will it cost to sequence a base pair of DNA in 2012, or what will the spatial and temporal resolution of brain scanning be in 2014, it turns out those things are remarkably predictable, and I'll show you a plethora of these logarithmic graphs with very smooth exponential growth. In the case of computing, it's a double exponential growth going back a century. And there is a theoretical reason for that, and we really can make predictions.

In terms of overall rate of technical progress, what I call the rate of paradigm shift, there's a doubling every decade. In the case of the power of information technology, bandwidth, price performance, capacity, the amount of information we're collecting, such as the amount of DNA data, the amount of data on the brain, the amount of information on the Internet, these kinds of measures double every year. But if we take the rate of technical progress doubling every decade to address your 1905 example, we made in the 20th century about 20 years of progress in terms of paradigm shift at today's rate of progress.

We'll make another 20 years of progress at today's rate of progress, equivalent to the whole 20th century, in 14 years. I agree with the idea that we are perhaps a century behind in terms of understanding biology compared to, say, computer science, but we will make a century of progress in terms of paradigm shift in another 14 years, because of this acceleration. This is not a vague estimate but one based on data-driven models. I have a group of 10 people that gathers data on just this kind of measurement, and it is surprising, but these models do have both a theoretical basis and an empirical basis. The last 50 years is not a good model for the last 50 years. At the *Time* magazine Future of Life conference, all the speakers were asked, "What will the next 50 years bring?" I would say all the speakers based their predictions on what we saw 50 years ago. [James] Watson himself said, "Well, in 50 years we'll see drugs that enable you to eat as much as you want and remain slim." But 5 to 10 years is a more realistic estimate for that, based on the fact that we already know essentially how to do it and have demonstrated that already in animals.

I agree with Rodney's point that the human brain and biology work on different principles. But these are principles we have already been applying. My own field of interest is pattern recognition. And we don't use logical analysis, we use self-organizing adaptive chaotic algorithms to do that kind of analysis, and we do have a methodology and a set of mathematics that governs that. As we're reverse-engineering the brain—and that process is proceeding exponentially—we're getting more powerful models that we can add to our AI toolkit.

And finally I'd respond to the question, How is it that one can make predictions about something so chaotic? Each step of progress in fields like computer science and biology is made up of many tens of thousands, hundreds of thousands of projects, each of which is unpredictable, each of which is chaotic. Out of this entire chaotic behavior, it's remarkable that we can make reliable predictions, but there are other examples in science where that is the case. Predicting the movement of one molecule in a gas is of course hopelessly complex, and yet the entire gas, made up of trillions of trillions of molecules, each of which is chaotic and unpredictable, has very predictable properties that you can predict according to the laws of thermodynamics. We see a similar phenomenon in an evolutionary process—biology was an evolutionary process, and technology was an evolutionary process.

VENTER: I find it fun to sit here as a biologist and listen. I agree with a great deal of it. There's a big difference though when we're talking about engineering single-cell species versus trying to engineer more complex ones. We have 100 trillion cells—when you try to use the same genetic code to get the same experiment done twice, it never works. So-called identical twins don't have the same fingerprints or footprints or the same brain wiring, because there are so many random events that creep in each time there's a cell division or some of these biological processes. It's pretty sloppy engineering if it's engineering. You can't get the same answer twice.

But with single cells, with bacterial cells, I agree they'll be the power plants of the future. It will happen even faster than either one of you are predicting, because there's the threshold now where we are trying to design a robot to build these chromosomes and build these species, that we could maybe make a million of them a day—because there are so many unknown genes that we basically have to do this in an empirical fashion and then screen for activities. Everybody's worrying about the Andromeda Strain sort of approach, but that's the way biology's going to progress, much more rapidly than we've seen it in this linear fashion.

BROOKS: Yes, that's an interesting point. I hate to be a naysayer, but Ray forces me to do it.

The idea of building a million of them and then assaying them and maybe having evolution happen *in situ* is the way we are going to speed stuff up. On the computational side, about 15 years ago we thought that evolution in silicon was going to take off and there was a lot of excitement about that. But we haven't quite figured it out. We're missing something. There's been 15 years of slow progress in the artificial-life field, but not the take-off we were thinking would happen in the early days of artificial life—at the Santa Fe conferences, and when the Santa Fe Institute was founded, back in the late '80s and into the early '90s.

Jack Szostak and other people have been doing real evolution in test tubes, because that's the sloppy way we know how to do it now—it may be, and this is sort of an advance which could change things, like the development of quantum mechanics, which totally rewrote physics—maybe somebody at some point will come up with understanding how to do evolution better in silicone and understand what we're missing. That will then lead to all sorts of fast progress. While I agree with statistical analysis of the future history, there are these singular events that we can't predict which will have massive influence on the way things go.

KURZWEIL: I can tell you what's missing, which is a real understanding of biology, and I said before, we're at the early stages of that. We have had self-organizing paradigms like genetic algorithms and neural nets, Markov models, but they're primitive models, if you can even call them that, of biology. We haven't had the tools to be able to examine biology. We do have the tools now to ' see how biology works, we have the sequenced genome, we're beginning to understand how these information processes work, we're being able to see inside the brain and develop more powerful models from that reverse-engineering process. Then the question that comes up is, how complex is biology?

I certainly wouldn't argue that it's simple, but I would argue that it's a complexity we can manage, and the complexity appears to be greater than it is. If you look inside the brain, look at the cerebellum, for example, and the massive variety of wiring patterns of these neurons which comprises half the neurons in the brain, yet there are actually very few genes involved in wiring the cerebellum. It turns out that the genome says, "Well, OK, there are these four neuron types, they're wired like this, now repeat this several billion times, add a certain amount of randomness for each repetition within these constraints"—so it's a fairly simple algorithm, with a stochastic random component, that comprises this intricate wiring pattern. A key question is, How much information is in the genome? Well, there are 3 billion rungs, 6 billion bits, that's 800 million bytes—2 percent of that roughly codes for a protein, so that's 16 million bytes that describe the actual genes.

The rest of it used to be called junk DNA; we realize now it's not exactly junk, it does control gene expression, however it's extremely sloppily coded, there are massive redundancies—one sequence called ALU is repeated 300,000 times. If you take out the redundancy, estimates of compression are that you can achieve at least 90 percent compression, and then you still get something that's inefficiently coded and has a low algorithmic content. I have an analysis showing there are about 30 million to 100 million bytes of meaningful information in the genome. That's not simple, but it's a level of complexity we can handle. We have to finish reverse-engineering it—but that's proceeding at an exponential pace, and we can show the progress we're making. It's kind of where the genome project was a decade ago.

BROOKS: Ray, you're going to make me a naysayer again. I hate this, but see these—they're cell phones, right? Ray's worked in image recognition, pattern recognition, I've worked in it, we can't get our object-recognition systems—which is just reverse-engineering that piddly little 16 megabytes of coding for the brain, or whatever it is, to be able to recognize across classes like a two-year-old can.

In 1966 at the AI lab there was a summer vision project, run by an undergraduate, Gerry Sussman, that tried to solve that problem. My own Ph.D thesis was in that area in 1981. We still can't do generic object recognition today in 2005, and people don't even work on the problem anymore, because you couldn't get funding for generic object recognition, because it's been proved so many times that it can't be done.

Instead, people work on specific medical imaging or faces. The generic object recognition problem is a hard problem. It's not just a matter of turning a crank on a genome for us to understand how that works in the brain. In the original proposal, written in 1966, Seymour Papert predicted we would gain the insight to be able to do generic object recognition. But it didn't happen then and it still hasn't happened. Some of these things, these singular sorts of events, we can't predict, and we can't just turn the crank on genes to get a deep understanding of what's going on in the brain, whether computational or transcomputational.

KURZWEIL: There's any number of bad predictions that futurists, or would-be futurists, have made over the decades. I won't take responsibility for these other predictions. But all you're saying, which is basically coincident with what I'm saying, is that we haven't yet reverse-engineered the brain, and we haven't reverse-engineered biology, but we are in the process of doing that.

We haven't had the tools to look inside the brain. You and I work in AI; we've gotten relatively little benefit from reverse-engineering the brain in neuroscience. We're getting a little more now. Imagine I gave you a computer and said, "Reverse-engineer this," and all you could do was put crude magnetic sensors outside the box. You'd develop a very crude theory of how that computer worked.

BROOKS: Especially if you didn't have a notion of computation ahead of time.

KURZWEIL: Yes, you wouldn't have an instruction set—you wouldn't even know it had an instruction set or an op code or anything like that. But you'd say, "What I really want to do is place specific sensors on each individual signal and track them at very high speeds"—then you could reverse-engineer it; that's exactly what electrical engineers do when they reverse-engineer a competitive product.

Just in the last two years, we're now getting the tools that allow us to see individual interneuronal fibers, and we can track them at very high speed and in real time. There's a new scanning technology at the University of Pennsylvania that can see individual interneuronal fibers *in vivo* in clusters of very large numbers of neurons and track them signaling in real time, and a lot of data is being collected. And the data is being converted into models and simulations relatively quickly.

We can talk about what is the complexity of the brain and can we possibly manage that? My main point is that it's a complexity we can manage. But we are early in that process. The power of the tools is gaining exponentially, and this will result in expansion of our AI toolkit and will provide the kind of methods you're talking about. But just because we haven't done it today doesn't mean we're not going to get there. We just have the tools now to get there for the first time.

BROOKS: I absolutely agree with you that we'll get there, but I question the certainty of the timing.

QUESTION FROM THE AUDIENCE: Can you talk about computation and the brain?

BROOKS: A long time ago the brain was a hydrodynamic system. Then the brain became a steam engine. When I was a kid, the brain was a telephone switching network. Then it became a digital computer. And then the brain became a massively parallel digital computer. About two or three years ago I was giving a talk and someone got up in the audience and asked a question I'd been waiting for—he said, "But isn't the brain just like the World Wide Web?"

The brain is always—has always been—modeled after our most complex technology. We weren't right when we thought it was a steam engine. I suspect we're still not right in thinking of it in purely computational terms, because my gut feeling is there's going to be another way of talking about things which will subsume computation but will also subsume a lot of other physical stuff that happens.

When you get a bunch of particles and they minimize the energy of the system, then when you get 1,000 times more particles, it doesn't take 1,000 times longer; it's not linear, it's not constant (there's a little bit of thermal stuff going on), but it's nothing like any computational process we can use to describe what happens in that minimization of energy. We're going to get to something, which encompasses computation, encompasses other physical phenomena, and maybe includes quantum phenomena, and will be a different way of thinking about what we currently call computation. That's what will become the new model for the brain, and then we'll make even more progress in knowing where to put the probes and what those probes mean, which currently we don't know too well.

KURZWEIL: Let me address it in terms of what we've done so far: Doug Hofstadter wonders, Are we intelligent enough to understand our own intelligence? Implying that he doesn't think so. And if we were more intelligent and therefore able to understand it, well, then our brains would necessarily be that much more complicated and we'd never catch up with it. However, there are some regions of the brain, a couple dozen, for which we actually have a fair amount of data, and we've been able to develop mathematical models as to how these regions work.

Lloyd Watts has a model of 15 regions of the auditory system, and there's a model and simulation of the cerebellum and several other regions. We can apply, for example, psycho-acoustic tests to Watts's simulation of these 15 regions, get very similar results as we get applying psycho-acoustic tests to human auditory perception. It doesn't prove it's a perfect model, but it does show that it's in the right direction. The point is, these models will be expressible in mathematics. Then we can implement simulations of mathematics in computers. That's not to say that the brain is a computer, but the computer is a very powerful system to implement any mathematical model. Ultimately that will be the language of these models.

QUESTION FROM THE AUDIENCE: History is full of wars and many other unpredictable events. There also appears to be a growing anti-technology movement. Don't these phenomena affect the pace of progress you're talking about?

KURZWEIL: It might look that way if you look at specific events, but if you look at the progress, for example, of computation, for which we have a very good track record through the 20th century—which was certainly a tumultuous time, with two world wars and a major depression in the United States and so on—we see a smooth exponential growth, double exponential growth. You see a slight dip during the Depression, a slight acceleration during World War II.

Very little of the population, historically, has been involved with the progress hundreds of years ago. Just a few people—Newton, Darwin—were advancing scientific knowledge. We still have strong reactionary forces, but a more substantial portion of the population is applying its intellectual power to these problems and to advancing progress, and it's amplified by our technology; we routinely do things that would be impossible without our technology. Rodney mentioned the role of powerful computers and software in doing the genome project.

The kind of social reaction we see now—Luddite reactions, reflexive anti-technology movements, and so on—are really part of the process; they don't slow things down. Even stem cell research is continuing—some people identify stem cell as comprising all of biotechnology, but it's really just one methodology and even it is continuing. These social controversies tend to be like stones in a river; the flow of progress goes around them. And if you track the progress in these fields and measure it in different ways, with dozens of different types of measures, you see a smooth exponential process.

QUESTION: Was there exponential progress in technology hundreds or thousands of years ago?

KURZWEIL: It would be hard to track them, because technological progress was so slow then, but I do have a chart I'll show you that goes back a long time in terms of the whole pace of biological and technological evolution, and it has been a continually accelerating process, and I do postulate that it's a fundamental property of an evolutionary process. The idea of progress now is deeply rooted, despite the fact that there are people who don't believe in it and movements that object to it. It's so deeply rooted that it's an evolutionary process at this point.

VENTER: It's a good point, and maybe it doesn't affect the overall predictions, but it certainly affects reality: There are an awful lot of people who don't go into stem cell research right now; a lot has been shut down, a lot of scientists have had to leave the country to continue their research, a lot of money that could have gone to it has been diverted.

It's probably the single most important area, if we're going to ever understand how our brains got wired. If we don't understand how stem cells work, we'll never understand complex biology beyond single-cell organisms. We saw the same thing with synthetic biology when we made the Phi X 174 virus and injected the DNA into *E. coli* and had it start producing viral particles just driven by the synthetic DNA. There ensued a huge debate within the U.S. government of whether to classify our research and shut us down and not enable us to publish our data because it might enable bioterrorism.

KURZWEIL: Just to clarify, I do strongly endorse a free system, and I'm opposed to any constraints on stem cell research. Rodney and I work in a field that has no certification of practitioners; your software developers aren't licensed software developers, and there's no certification of products, despite the fact that software is deeply influential. My feeling is that we don't balance risks appropriately in the biological field, and people are concerned now that the FDA should be more strict, because nobody wants drugs that are going to harm people. On the other hand, we have to put on the scale the effect of delaying things.

If we delay stem cell research, gene therapy, some heart disease drug, by a year, how many hundreds of thousands of lives will be disrupted or destroyed by that delay? That is very rarely considered, because if you approve a drug and then it turns out to be a mistake, a lot of attention goes on that. If something is delayed, nobody really seems to care, politically. We should move toward an open system—there are down sides to all of these technologies, there are risks. Bioterrorism is a concern, not just using well-known bioterrorism agents but creating new ones. But we make the dangers worse: A bioterrorist doesn't have to put his invention through the FDA, but scientists like Craig, who are working to defend us, are hampered by these regulations.

BROOKS: Let me follow up on Craig's point. Scientists are being driven offshore—but also we have the problem that because of the September 11 event, we are having trouble getting foreign students into our universities, and that's really slowed things down. The students who are here are still scared of going home or going to a conference outside the country, and we can't get as many students as we could before, and that's having a real impact, and it's slowing a lot of things down.

VENTER: The immigration issue has been flagged as the number-one issue at the National Academy of Sciences affecting the future quality of science and medicine in this country.

KURZWEIL: You'll be pleased to know that things are not slowed down in China and India. I've got some graphs showing engineers' levels are declining in the United States; they're soaring in China. China's graduating 300,000 engineers a year compared to 50,000 in the United States.

QUESTION: What do you see happening in the immediate future?

KURZWEIL: There are lots of things that are not predictable. The attributes representing the power of information technology turn out to be predictable. It's a chaotic process, but technology, and certainly information technology, proceeded smoothly through World War II, despite the fact that it was a very destructive time. We don't know what the future will bring.

These technologies aren't necessarily beneficial. Everybody in this room is trying to apply these capabilities to promote human values and overcome disease and so on, but it would only take a few individuals to cause a lot of destruction. We don't know how these technologies will be applied. We can discuss how best to allow creative projects that will advance human knowledge and reduce human suffering, but the future hasn't been written, even if certain attributes of information technology are predictable.

QUESTION: One of the important attributes of biological systems is that there's tremendous plasticity. I'm wondering what you see as the most compelling engineering exemplar of the systems that have some sense of plasticity. We look at dynamic adaptation or things of that kind, neural networks. Really serious computation brings us to computer science. Where do you see that going?

KURZWEIL: We're in the early stages of a great trend, which is to apply biologically inspired models. As we learn more precisely how biology works, we'll have more powerful paradigms. But there are a lot of self-healing systems that are adaptive—there's been a lot of progress in the last five years on 3-dimensional molecular electronics, which is still formative, but these circuits will have to be self-organizing and self-healing and self-correcting, because if you have trillions of components, you can't have one misplaced wire or one blown fuse destroy the entire 3-dimensional mechanism.

Even circuits on the market today that are nominally flat have so many components that they're beginning to incorporate these self-healing mechanisms where they can route information around areas that are not functioning. The Internet itself does that, and as we get to the World Wide Mesh concept, it will become even more self-organizing. Right now all of these little devices are spokes into the Internet, they're not nodes on the Internet.

But we're going to move to where every single device is a node on the Internet. So in addition to allowing me to send or receive messages while I'm sitting here, this phone will be transmitting and forwarding messages from other people. I will be a node on the Internet, and my phone will allow that to happen, because I will also be taking advantage of this mesh capability. We are moving toward much more of a self-organizing self-healing paradigm. IBM has a big project on self-healing software to manage IT networks. There are a lot of examples of that.

VENTER: It's a critical point. As part of trying to make synthetic life, we're having to come up with a definition of what is life, and one of the key components we've found in every genome we've done is a gene called REC-A, involved in DNA repair. Repairing DNA is one of the most fundamental components of life, but an even more fundamental component we've seen in every species we've determined is built-in mechanisms for continued evolution. The plasticity we see in even the simplest sets of genes in a single cell—that's in fact why we're having to build a cell from scratch, because we can't determine empirically what genes cover up for the functions of others—but every genome we've done has built into the DNA mechanism some of these remarkably simplistic things.

And *Haemophilus influenzae*—every one of us in this room has a different strain of it in our airways, because it continually goes through Darwinian evolution in real time. There are these tetrameric repeats—four-base repeats—in front of all the genes associated with the cell surface proteins and lipoproteins. And every 10,000 or so replications, the preliminary slips on these—that's called slip-strand mispairing—and it shifts the reading frame and the genes downstream, basically knocking them out.

Just by knocking out genes in a random fashion, it constantly changes what's on the cell surface, and that's why our immune system can't ever catch up. It's constantly winning the war against our immune system, and we have basically millions of organisms in us that are adapting in real time. The complexity hasn't even begun to be approached in terms of what biological systems can and really do.

QUESTION: What about synthetic genes?

VENTER: It's an issue we're facing: whether or not we have synthetic genomics. We talked earlier about the influenza virus; you get two of these viruses in one animal and they can recombine to form all kinds of new molecules that put—if it's like the last pandemic—75 million people at risk for death very rapidly. There's little difference between a new emerging infection and a deliberately-made pathogen, in terms of the impact it has on humanity. We need defenses against them. I've argued that it's never been a waste of the government's money to try to develop new antivirals or new antibiotics or new approaches to treat these infections across the board, whether or not there's ever a bioterrorism event.

The new technology we've developed, wherein we can synthesize a virus in two weeks, does have some clear implications if someone really wanted to do harm through these techniques. Some of the arguments around it are that it's a lot easier to obtain any kind of lethal organism you want through much simpler means than trying to synthesize it. All the efforts that went on to try and change smallpox and anthrax were major state-supported events in the U.S. and the former Soviet Union.

These are not simple processes right now. We could be in the place in ten years where you could be the first one on the block that builds your own species in a garage, but we're not quite there yet, and at the same time these events—and we get literally thousands or millions of human-made organisms—increase the chances of getting new approaches for counteracting them. On the engineering side, it's easy to build into them mechanisms so that they can't self-evolve, so that they can't survive if they escape from the laboratory. But with these same techniques we've developed it wouldn't be difficult for our group to build a smallpox virus based on the DNA sequence in a month or two.

KURZWEIL: There are a couple of characteristics that affect the danger of a new virus; obviously how deadly it is and how easily it spreads. But probably the most important is the stealthiness. SARS spread pretty easily, and it's pretty deadly, but it's not that stealthy, because the incubation period is fairly short. New naturally emerging viruses don't tend to have the worst characteristics on all of these dimensions, so one could, if one were pathologically minded, try to design something at the extreme end of these various spectrums.

I did testify before Congress recently advocating that we greatly accelerate the development of these defensive technologies. It's true that it's not easy to create an engineered biological virus, but the tools and the knowledge and skills to create such a bioengineered pathogen is more widespread than the tools and the knowledge to create, say, an atomic bomb and could potentially be more dangerous. We are pretty close at hand to some exciting broad-spectrum antiviral techniques. We could apply, for example, RNA interference and other emerging techniques to provide an effective defensive system. It's a race: We want to make sure we have an effective defense when we need it. Unfortunately the political side doesn't get galvanized unless there's some incident. Hopefully we can interest the funding sources before it's needed.

14
Engineering Biology

Drew Endy

[February 17, 2008]

Drew Endy is a professor of bioengineering at Stanford University.

How can I make biology easy to engineer? Going back hundreds of years, people imagined that you could always design and build or make life, but nobody could do that much about it. As the 1970s rolled out, human beings invented a lot of technology: recombinant DNA for cutting and pasting pre-existing fragments of genetic material; the polymerase chain reaction, invented in the '70s but not really figured out until the '80s; automatic sequencing, with Fred Sanger in 1977.

Now, 30 years after the initial successes of biotechnology, it has realized only one of the early promises. The early promises were: first, making therapeutics via recombinant organisms, producing drugs like insulin via bacteria, which has worked. Gene therapy was the second promise—fixing genetic defects by patching our DNA, and this has not yet worked. And third, to develop crops that could fix nitrogen, so that agriculture wouldn't have to rely on synthetic fertilizers. That hasn't worked either. So, of the three great early promises that were rolled out with the beginning of genetic engineering, we have realized one of them.

Nevertheless, biotechnology exists. It's a huge positive contributor to our health and economy and the human condition generally. So the question is, Can we realize the initial promise of biotechnology? Or, forget that question: How do we make biology easy to engineer, so that anything we might want to manufacture out of the living world is something we can pull off?

Imagine you're fifteen or seventeen or eighteen years old. You're an ambitious youngster, and you're showing up as a first-year undergraduate, and you're choosing what to major in. Well, you could choose to major in biology or electrical engineering or computer science or . . . Oh, now you can major in biological engineering! What would you expect to learn? What would you expect of your faculty colleagues, your professors? What would they be able to teach you?

You look to your friends who will study electrical engineering; they can learn how to design and build computers, or write computer programs, and the objects they make don't have emergent properties unless that's what's intended; instead, they behave as expected. Then you look at biological engineering and you say, "Well, yes, I'd like to design and build living organisms, or program DNA to execute genetic programs that behave as expected." But nobody can teach you how to do that.

Thirty years into biotechnology, despite all the successes and attention and hype, we still are inept when it comes to engineering the living world. We haven't scratched the surface, and so the big question for me is, How do we make biology easy to engineer? For comparison, take modern electronics: During and following World War II, people were building computers. John von Neumann was building a nice machine in the basement of the Institute for Advanced Study at Princeton. The official purpose of this machine was to design hydrogen bombs and compute the trajectories of munitions. And he, of course, is apparently running artificial-life programs on it, because that's what he's more interested in. Let's say it's 1950. The Apple-1, the personal computer is only 25 years later.

Will we ever get to the point where biotechnology is not an exclusive technology, not a technology that requires experts? Will we ever get to the point where we can make many-component integrated systems? Will we ever get to the point where we have separation of the types of work in biological engineering, so that one person might be an expert designer, another might be an expert constructor, as we have expert architects and builders and what not?

And a parallel question is, What are the consequences of success? If you look around the room we're in, everything in the room is a synthetic or engineered artifact, right? Even the air we're breathing has been engineered for temperature and humidity. The only thing that hasn't been engineered are the living things—ourselves. Again, what's the consequence of doing that at scale? Biotechnology is 30 years old; it's a young adult. Most of the work is still to come, but how do we actually do it? Let's not talk about it, let's actually go do it, and then let's deal with the consequences, in terms of how this is going to change us; how the bio-security framework needs to recognize that it's not going to be nation-state-driven work necessarily; how an ownership, sharing, and innovation framework needs to be developed that moves beyond patent-based intellectual property and recognizes that the information defining the genetic material will be more important than the stuff itself and so you might transition away from patents to copyright; and so on and so forth.

So, to zoom out, how to make biology easy to engineer? And how do we do this in a way that leads to constructive culture around the technologies that's overwhelmingly positive in terms of the consequences of its being rolled out?

What happens when the technology in support of engineering biology is sufficiently advanced that somebody like Stefan Sagmeister, the graphic designer, could sit down and design a life-form he would consider interesting or beautiful? How do we get from what we've got today, where we're basically celebrating a bunch of stunts and we've delivered only a third of the initial set of promises of biotechnology and there are so many other things we can imagine that are fantastical because they're just too complicated given the current state of affairs? How do we get from that to "Yeah, the graphic designers are making beautiful living objects?"

There is also the issue of addressing energy needs. A lot of people drive investments in biotechnology from the application side, and that's good. There are lots of pressing human needs and problems: Food, which is an energy of sorts for people and animals. Liquid fuels for cars and jets, and then you've got health and medicine, and then you've got environmental issues, and then you've got materials construction, and ta-da-da. What's interesting about biotech is that the applications have always been so unbelievably pressing.

So let's wind the clock back 30 years; to a first approximation, there's been an under-investment in tools, right? Say you're running a team that's trying to figure out how to make insulin in bacteria, or how to make artemisinic acid for treating malaria in bacteria or yeast, and somebody says, "Hey, why don't you take 5 percent of your project's budget and, instead of spending it on delivering your product as fast as you possibly can, leave behind a little bit of engineering infrastructure, so that the next time you do a project like this it doesn't cost you $40 million? So that the next time you do a project like this, it's much, much easier?" The arguments in response to those sorts of suggestions are: If we delay shipping our product by a day, we will lose to our competitor, or 10,000 additional people will die, or something. On a short time scale, it's impossible to argue against these positions, right?

But if you take a longer view, absent making such foundational investments in technologies that support the engineering of biology, the engineering of biology will always be hard. We have to figure out how to solve that problem. This comes into play when you think about energy. What do I think about the biological production of energy? Terrific, right? Yeah, we'd rather not be burning dinosaur juice. It seems like important work, and the lab next to mine, when I was earning my Ph.D, was working on cellulosic ethanol. If the price of oil went up by a factor of 2, cellulosic ethanol would be cost-competitive. That was in 1994.

I hope bioenergy succeeds. But consider that there was a trap for John von Neumann when he was building those early computers to compute the trajectories of munitions. It turns out that the utility of computers is much more than we could have imagined, much more than the military applications and accounting databases. But with a few exceptions, nobody back then had a clue what those applications would be. Thus I'm not interested in pursuing any one application in biotechnology right now, because I want them all to come true, and I want them all to come true on a time scale relevant to me—I can be very direct and selfish about this. And the only way that will ever happen is if I don't go work on bio-energy. There are enough people who'll work on that, because it's a problem everybody can understand; you'll be able to raise resources around it and go do the work. There's the complementary problem, the meta-level problem, which is, Let's make all of biotechnology easier for everybody.

The underlying goal of synthetic biology is to make biology easy to engineer. What does that mean? It means that when I want to build some new biotechnology, whether it makes food I can eat or a biofuel I can use in my vehicle, or I have some disease I want to cure, I don't want that project to be a research project. I want it to be an engineering project. In the science of biology, the people you're talking to are scientists, they're not engineers, and—not to be arrogant, just to be an observationalist—the question is, If you're an engineer looking at biotechnology, what do you need to do in order to make it easy to engineer? That's what synthetic biology is about.

You could start talking about historical examples of what engineers do when faced with situations of this sort: In America, in 1860, machinists built objects—steam engines, what have you. Nuts and bolts that held together machines were specific to the particular machine shop that manufactured them. What that means is, if you buy a machine from a machine shop in Newark, New Jersey, and it breaks down in Chicago, you have to send it back probably to that specific machine shop, where the machines are set to tool things on a particular set of designs, in order to get the replacement part or get the thing fixed.

In April of 1864, somebody said, "Enough!" William Sellers, of the Franklin Institute in Philadelphia, gave a paper on a system for nuts and bolts. And he proposed the Sellers Screw Thread standard, which is a 60-degree angle squared off at the top screw-thread design, easier to manufacture than the English Whitworth standard of a 55-degree angle, rounded screw thread. As a result, eventually everybody in the U.S. retooled their machine shop to produce screws, nuts, and bolts in accordance with the Sellers standard. The consequence of this today is that when I go to the hardware store and get a nut and a bolt, so long as they don't screw up the English/metric thing, I can take those two objects and put them together.

That's an example of what an engineer would call reliable physical composition. Take two objects and put them together. The other thing that happens is that when you have the nut and the bolt together as a composite object, when you pull on the nut it stays put. It doesn't come flying off. The composite object has the expected behavior; it doesn't have some emergent property. That's reliable functional composition. The function of the two things when you put them together is what you'd expect. What's amazing is that I've taken this standard for granted my whole life. Even though I have three engineering degrees, I didn't know about this until a couple of years ago, when Tom Knight of MIT pointed out to me that it would be nice if we had standard biological parts that could snap together and behave as expected when we did.

George Church has been exposed to this, but it's not of his mother culture. He's a geneticist; he's reverse-engineering natural biological complexity. That's a great thing to be doing. Engineers hate complexity. I hate emergent properties. I like simplicity. I don't want the plane I take tomorrow to have some emergent property while it's flying. If you look at the science of genetics, which has been in the business of trying to figure out the relevant information in coded DNA, the most important thing has been technology. Before DNA sequencing existed, people would find mutations and they'd map them, in the process, to different regions on the DNA. And the mathematics that was being used was based on simple logic. Then a number of great people drove forward with the sequencing of DNA, and as a result of that technology we can now read DNA, and that technology continues to get better.

It's important to put the impact of advances in DNA-sequencing technology in context. In 1990, nobody had sequenced anything except for a couple of bacterial viruses and maybe some other viruses. In 1995, the first bacterial genome, *Haemophilus influenzae,* was sequenced. In 2001, there's a draft of the human genome sequence. How did we, in the 1990s, go from stinking at sequencing DNA to "Yeah, we just sequenced human beings," and now, only seven years later, the personal-genome projects coming online? It's not because George Church and Craig Venter and Eric Lander and Francis Collins got 10 billion times smarter during the Clinton years. It's because the technology for sequencing DNA got automated and scaled up sufficiently to do it.

The impact of underlying technologies on what's possible is an important thing to recognize, and something that, if successful, you want to be able to ignore in the same way I want to be able to ignore nuts and bolts and screw-thread standards. Genetics changed in response to sequencing technology. You could read DNA. We have no idea what it says. The mathematics is now pattern recognition, to try and look at many sequences and find the conserved patterns that might have relevant functions. Synthesis technology is coming online next.

2008 is our 1995, if you will. This is the year when a bacterial genome was synthesized from scratch. Ahead of that work, chloroplast genomes, mitochondrial genomes, were constructed; in fact, a project from Japan a couple of years ago made a 10-million-base-pair fragment of DNA from existing fragments, which is 15 times larger than anything getting attention these days.

So, what happens to the science of genetics as a new set of tools come online that lets us build whatever DNA molecule we want, and you get to make changes and see what happens? Instead of being called genetics, this is called reverse genetics, and the mathematics driving this is probably going to be perturbation design. What changes do you want to make, and how do you choose what to make? First, genetics goes from pre-sequencing technology and it's based on logic. Then it's post-sequencing and it's pattern recognition. And next there's going to be post-synthesis genetics, and it's going to be, "Make whatever you want." Perturbation design becomes the mathematics. And the whole field's going to change.

When sequencing technology was developed, the scientific community—not to mention the rest of the world—did an incredibly poor job of anticipating the resulting challenge of, "What the heck does all this DNA-sequence information mean? How big is the pattern recognition problem?" Fields of science like bio-informatics are purely reactionary and have poorly planned responses to technology advances, and we're going to get the same thing again with synthesis.

For example, how do you manage the information going into a DNA synthesizer so that you can construct some useful object that will help you do genetics? This is the reverse bio-informatics problem. George Church and Craig Venter have a lot to contribute to it, which will be terrific. It will be part of synthetic biology, but it will be synthetic biology impacting science, which is the worst-case scenario for synthetic biology. We fail to actually deliver any useful artifacts that people want, but at least we'll fail, and we'll de-bug our failures, which will prioritize our misunderstandings of biology much more ruthlessly than anything else, and which is much better than a NIH study section.

What else might happen? I've got an invitation to give a talk at the Chaos Communication Congress, which is the largest hacker meeting in Europe, about 4,000 people—people who like to make stuff, people who like to understand how things work. And they're very interested in learning how to program DNA and how DNA works. One consequence of actually making biology easier to engineer, whether you're standardizing the components or figuring out how to develop higher-level programming language, is that other people besides the usual suspects are going to have access to the technology.

If you think about what happened between 1950 and 1975, when you went from von Neumann's machine to the Apple-1, a key part of this transition was that folks were so stoked about computing and so fed up with limited access to centralized computing resources that they went out and built their own computers—by definition, the personal computer. As a result, today we have a worldwide community of folks who are excited about building electronics and writing software, which includes school kids, professionals, big companies, small companies, governments, you name it—a diverse ecology around that technology.

Programming DNA is more cool, it's more appealing, it's more powerful than silicon. You have an actual living, reproducing machine; it's nanotechnology that works. It's not some Drexlerian fantasy. And we get to program it. And it's a pretty cheap technology. You don't need a FabLab, like you need for silicon wafers. You grow some stuff in sugar water with a little bit of nutrients. My read on the world is that there is tremendous pressure, just starting to be revealed, around what heretofore has been extraordinarily limited access to biotechnology.

Take some of the writings of Freeman Dyson. He's imagining genetic engineers of the future winning the Philadelphia Flower Show, the San Diego Reptile Show, or whatever it is. How do you get there? And as you start working through that path of getting there, what you find is that there are vast communities of people who want to be doing this. But the people promoting the technology tend to favor exclusive ownership and limited access and present themselves as godlike creators—as opposed to

"We're constructing things; we could use your help; anything we do today is going to pale in comparison to what's coming, so let's figure out how to work together on this."

As a different example: In 2003, I taught a course at the Synthetic Biology Lab at MIT with some colleagues, and we had 16 students. For the last four years, this course has been doubling every year, and it's now taught independently at about 60 schools in 30 or 40 countries worldwide. It's called IGEM, the International Genetically Engineered Machines competition. There are teams of teenagers from Germany programming DNA happily there, as well as in Australia, Russia, Japan, China. The competition was won by the team from Peking University this year, and 600 or 700 students participated.

How do you recognize this potential and serve it and bring more people to participate in it? The rewards of doing so are greater than any one group's project. For instance, the team from Melbourne, Australia, showed up with a 6,000-base-pair fragment of DNA they found, which somehow—I don't know how this actually works—folds up; the proteins get made and the proteins self-assemble into a 50-nanometer sphere filled with gas. The protein shell is somehow gas-impermeable, and these little balloons, these protein balloons, get booted up inside the cytoplasm of cells, and you can control how many different balloons there are. Depending on the number of balloons, the cells will either float or sink or be neutral.

Who knew? I didn't know anything about this biology, and they showed up, they made this standard biological part, such that we can now snap it together with the 2,000 other parts we've got in our collection so far, which is a free collection. We shipped over 100,000 parts around the world last year for free, and the collection's doubling in size every year.

If you make biology easy to engineer, and you make it accessible, by definition people will learn about it. You can talk to computer-programming conferences about it. It's a very different world from going around claiming you've created life. It's a very different world from going around filing patent applications that say you've invented the idea of a synthetic genome. It's a very different world from spending $40 billion on a classified biological defense facility at the site of the past U.S. Offensive Biological Weapons program. And so there's a cultural mismatch.

The mismatch is largely generational, and it's also largely perspective-driven. By that I mean that the previous generation of people working in biotechnology were scientists, and the ones coming up now are engineers. We'll have to invent our new world of biotechnology, and I suspect we'll learn lessons on biological safety from the past generation, but all the other lessons are up for grabs. The bio-security framework will collapse. The IT framework based on patents won't scale, and the questions of playing God or not are so superficial and embarrassingly simple that they won't be useful in discussion.

The more serious situation is that these issues of human practice don't get resolved in a six-month conversation. It's not like what happened in Cambridge, Massachusetts, in the '70s, when recombinant DNA work got shut down for a bit and then became OK. The technologies are being developed and distributed so quickly, yet there's still so much more to do in improving the work of biological engineering. The conversations we need to set up are conversations that need to persist in constructive ways for decades.

The open-source world is one thing; if you're trying to invent a language for programming DNA, having a proprietary language seems stupid. If Oxford University had supported privatization of the English language hundreds of years ago, the dictionary they made wouldn't have been so useful. There will be a core collection of standardized genetic objects that can define families of languages people can use to program DNA. And those have to be made a public resource.

This will be a big transition from today. Biotechnology today derives investments from business models that support the exclusive application of different biological functions for a small number of problems. For example, there are wonderful companies that have locked up most of the relevant intellectual property around how to engineer proteins to bind DNA. The products that they can deliver will be measured in small positive integer numbers, a few diseases.

But the real value associated with being able to engineer proteins that bind DNA are in the uncountable applications people could use the proteins for. It's like a programming language where it would be a big downstream economic cost if you owned "if / then" and you were the only person who could use it. We need to be able to reuse this stuff in combination. Note that the ownership of biotechnology will play out in a landscape that's surfing along a technology transition where, as automatic construction of DNA gets better and better and better, you'll care less about the specific material you have, you'll care more about the information on a computer database and the computer-design tool that lets you organize that information, compile it down to a DNA sequence, and print it. As soon as you start to manage information, all sorts of new ownership, sharing, and innovation schemes become allowable.

Where will we be 30 years from now? 1995: *Haemophilus influenzae* sequence. 2001: draft of human genome available. 2007: multiple chromosomes assembled from scratch, bacterial virus, or organelle. By 2012, the design of eukaryotic chromosomes should be routine. Also, five years from now we may have just begun to make some good progress on reliable functional composition of standard biological parts. Nobody knows how expensive solving that problem will be, but because biology works, there are plenty of existence proofs. If I had to guess, I'd say we'll have a collection of tens of thousands of genetic objects that support reliable functional composition between 10 and 15 years from now.

OK, let's cut it a different way: I'd estimate the cost of synthesizing the DNA of every human being on the planet born in the next year at $10 trillion. That's 20 percent of the world's economy. That number is dropping by about a factor of 2 every 12 to 18 months. On what time scale does it become worth considering whether or not we can afford to construct every new human genome that will come into existence, and we can decouple the designs of human beings from the natural constraints of direct descent and replication with error? My sense is that technology will support this well in advance of our ability to have any conversation on the consequences of using the technology. It's not a 50-years-off thing, it's not a 30-years-off thing, it's probably not even 20 years off, in terms of where the technology needs to get.

I like to build stuff, and biology is the best technology we have for making stuff—trees, people, computing devices, food, chemicals, you name it. I somehow found my way to biology and, along with the ambition of getting better at engineering biology, there's this wonderful complementary puzzle of how the hell does this stuff work? All these living systems we inherit from evolution.

I was fortunate in the early '90s to find an engineer, John Yin, now at Wisconsin, who knew something about DNA. He had just come back from working with Manfred Eigen in Germany, and was studying virus evolution. He was at Dartmouth College, so I did my Ph.D up there and had an interesting experience as an engineer trying to develop computer models to help biologists understand the architecture of the natural genetic systems they were changing. I had some hypotheses coming from my work, and I tried to get some biologists to do experiments for me; I was not successful in doing that. In hindsight, I recognize that that's because any good biologist who does experiments has multiple lifetimes of work to do. They're never going to do your experiment, so you need get in the lab and do it yourself.

That took me to Austin, Texas. At the University of Texas, I worked with Ian Molineux, who had done the early PCR work at MIT; he was running one of the last bacterial virus labs in the country, and he taught me how to map and clone DNA and do my experiments. I then spent a summer in Madison, Wisconsin, and then went to Berkeley, where I ended up working with Sydney Brenner and Roger Brent, two good biologists, in an independent not-for-profit; our mission was to do the next generation of biology, whatever made sense. Part of my work there included taking a look at my results from Texas, and I noticed that all of the predictions I had made using my computer models, about how these natural biological systems would behave when we changed them, turned out to be wrong, especially the interesting predictions. I would want one behavior, and when I went to make the change, exactly the opposite would happen. In this situation, engineers do what's called a failure analysis. So I made some predictions: "I'm going to make these changes to the architecture of this virus and as a result, the virus will grow faster." I would go into the lab, make those changes, and it would grow slower. My modeling tools weren't good enough to support purposeful determinative changes that would result in the behavior I expected.

This was a sufficiently painful process to give me a lot of time to think about why things weren't working out. And the conclusion I came to at Berkeley was that evolution is not selecting for designs of natural biological systems we can understand; the things we inherit from the living world have not been selected for ease of understanding, let alone ease of manipulation. It's not part of evolution's objective function.

If I wanted to be able to model biological systems, to predict their behavior when the environment or I made a change to them, I should be building the biological systems myself. That, for me, was the transition to what's now called synthetic biology. I started broadcasting that idea in the 1990s. The only person who returned a coherent signal intellectually was Tom Knight, in the electrical engineering department at MIT. Tom had self-started in biology five years earlier and is now, in addition to being one of the best engineers I've ever met, one of the best microbiologists I've ever met. Tom was interested in it from his own perspective, having mostly to do with building computers. We need to use biology not to be a computer but rather to build our computers, because we're going to need to put atoms exactly where we want; as semiconductor devices get smaller and smaller and smaller, you can't rely on random distribution of the dopant atoms in the devices. The statistics go to heck, and you have to figure out ways of putting exactly one or two or some small number of dopant atoms in every little gate you've got.

So that brought me to MIT, in January 2002. And along with Tom, there was the opportunity to be responsive to this bigger opportunity: Let's engineer biology. We're going to pull off a new department and a new venture in biological engineering. In the context of MIT, this was not the first time it had been tried.

There's a 1939 paper by Karl Compton, then the president of MIT, whose title is, "The Genesis of a Curriculum in Biological Engineering." It describes the ambitious and impressive five-year major, where you get a dual degree in biophysics and biological engineering.

Somehow that earlier effort crashed and burned. I haven't completed my failure analysis yet, and I don't know if it was simply World War II and the redirection of interests or other stuff, but what's interesting to consider is that 1939 is exactly when the Rockefeller Foundation was making investments in the science of biology, arguing, correctly, that the relevant physical level of resolution at which to understand the living world is atoms and molecules. That becomes molecular biology. And so biological engineering could have got started at that same time, but didn't. So that's how I get to where I am.

I've resigned my position at MIT; I'll be moving to Stanford next summer. It's a place that will support the scale of foundational research that needs to happen in biological engineering. There's a tremendous engineering community in the Bay Area, in electronics and software, and these are the folks who have the most relevant skills. If you look forward to what the challenges are in biological engineering, the main challenges are how to manage complexity, meaning how to produce simplicity in a many-component integrated system, and how to develop a theory that supports the programming of evolution.

And there's a way to think about this that maps directly onto communications theory, where you think about a sender and a receiver and a message being transmitted along a channel. In evolution, you have a parent generation, which is the sender, the transmitter, and you have the progeny, the children, which are the receiver. The message that's being transmitted is the design of the living organism, and the channel the signal's being propagated along is the process of replication of the machine. So, in any case, the work force and knowledge base most relevant to the future of biological engineering is now on the San Francisco Peninsula.

When we organized the 1st Synthetic Biology Conference at MIT in 2004, we were expecting about 150 people; and 500 people wanted to come, given 6 weeks of notice. Now it's going on 4 years later; the 4th meeting will be at the Hong Kong University of Science and Technology, which I think will blow doors off most places in the world 20 years from now. The university is in Clear Water Bay in Kowloon.

It's interesting for me to learn how difficult it is for folks to appreciate what an exponential technology implies. The fact that sequencing went from approximately zero to human genomes in 10 years. The same thing is happening with construction of genomes. And with the parts collection—the standard biological parts doubling every year. And the same thing is happening with the number of teenagers who'd like to do genetic engineering; it's doubling every year. How do you live in a world where you're surfing that exponential in a constructive and responsible way? Very few people get that.

15
Eat Me Before I Eat You: A New Foe for Bad Bugs

Kary Mullis

[March 17, 2010]

Kary Mullis received a Nobel Prize in chemistry in 1993 for his invention of the polymerase chain reaction (PCR).

We're working on a way to manipulate the existing immune system so it can attack things it's not already immune to. We've been controlling bacteria for years with antibiotics, but the bacteria are catching on. We've never been good at controlling viruses, unless we prepare for them well in advance by vaccination, but now we can use the same method for them, too—and in both cases the cure is not administered until you're infected and it works right away. It sounds too good to be true. So did antibiotics—they were called "miracle drugs."

In order to understand what we're doing, I should explain how the immune system works. Most people know that you've got this system but not how it functions on the level of molecules and cells. It's a collection of lots of different kinds of cells, each with their own purposes. There are about as many as you have in your brain, distributed mostly in special areas all over your body. The business end of the system is a set of hungry cells that will destroy and ingest things that are designated by the whole system as being "other." The rest of the system is charged with preventing them from eating anything else. New cells are always being born, and they are right away tested for their ability to make antibodies that attach themselves onto things that are "other". Antibodies are molecular markers.

If they make antibodies that attach to something that's "you", the immune cells are killed and also eaten. Not much goes to waste. There is a clever selection process under way. Right after an immune cell is born, a special part of its DNA is scrambled uniquely. The scrambling is done largely by an enzyme we picked up from a retrovirus that a life-form was infected by 60 million to 80 million years ago—something maybe slimy but with teeth, which was dreaming about becoming us in the long course of evolution. The virus managed to get into the genome of our germ plasma. About half our genome, by the way, is used genes, picked up from viruses. It's shocking and a little humbling to hear this. As a result of DNA scrambling, every new immune cell is genetically uniquely encoded, in one region, to produce a particular protein structure that has never been seen before and that hopefully will be able to bind to some structure which at the time is unknown.

If it makes a protein that binds to anything currently in your body, that "anything" is most likely going to be "you", and so the cell is killed. But if the protein, which is referred to now as a B-cell receptor, can't bind tightly to anything presently around, then the new cell is allowed to live. These young immune cells are escorted around your body, looking at various tissues to make sure there's nothing native in you that its B-cell receptor will bind hard to.

After that, it's just left alone, or it hangs out in a lymph node. If something foreign appears in your body that it *can* bind to, a reasonable assumption is that the thing is not "you". There are other cells that sense when there are increasing numbers of this foreign thing, indicating that it could be a threat, and an immune response gets under way. What began as a lone cell able to make an antibody resembling its unique B-cell receptor, except that the antibody can be excreted in quantity, is instructed to divide itself as fast as possible, and the daughter cells start pumping out the antibodies. The antibodies attached to the invader are an invitation for all kinds of specialized immune cells to have their way with it. The elapsed time since the foreigner arrived can have been weeks.

It took a long time for scientists to understand how, from a limited genome, we can make antibodies that bind tightly to anything at all. Something from Mars could show up in your body and you could make an antibody to it. It didn't make sense. There wouldn't be enough information in your DNA to make a strong binding site for all possible entities. Sir Frank Macfarlane Burnet came up with the explanation just offered. It generally works.

One of the problems is that the immune system might not figure out that there's a foreign entity present until that entity has already multiplied rapidly. A bacterium can reproduce itself every 30 minutes, so the numbers go up fast when you get infected with bacteria. If the right cells aren't there in the little spot where the bacterium is, it may take quite a while before your immune system responds.

This long bureaucratic process by which the right cell is put to work can be described as a hierarchy of immune cells having to make a lot of decisions. The question would be whether this thing is reproducing rapidly enough that we need to make a response at all. And if so, what kind of a response? And every single action your immune system takes causes some collateral damage. And so it's not a cheap thing to do, for one thing. It's a serious decision. It usually takes a couple of weeks before you have a really strong immune response to any particular new thing.

This immune response lasts until there are many antibody receptors on the growing clone that are left empty. When the system senses an abundance of these, it realizes the foreign thing is gone. It's like an army withdrawing from war. Most of those cells specifically suited to fight the defeated entity are slowly eliminated. A few of them are kept, like keeping a reserve. They're called memory cells, so if you ever run into that thing again, there are at least 1,000 cells to start with, instead of one. This way, you can make an immune response faster.

I started thinking about this and how we might help it along. It occurred to me that there are certain powerful immune responses we have from shortly after birth, and we keep them powerful, prepared to act at any time. They target things that are fairly common in our environment. One of these things is called the alpha-Gal epitope. It's a fairly simple trisaccharide that happens to be chemically synthesizable in a lab. About 1 percent of our immunity is devoted expressly to it.

What if you could chemically alter those antibodies with a drug, such that they would bind to something else? Something you had just contracted and you'd like to be immune to today? Instead of your immune system having to figure out what you have, a hospital lab could figure it out. Perhaps a patient has *Staphylococcus aureus*. Chemists could devise a linker molecule that, on one end, would bind to some part of Staph and on the other it would sport an alpha-Gal epitope. The alpha-Gal antibodies would bind to the alpha-Gal and thereby to the Staph. It's a clever trick and so far it works. It's applicable to any organism that has something specific on its surface, and all organisms do.

We can now easily look through the literature of the 10 or 20 different organisms that are starting to escape antibiotics, to determine what their surface proteins are. You always find some little fragment that stands out. Just as if you were looking at a person, you could say, "This person has a funny little ear, and he's always going to have it when I run into him. If I can get something to clamp to his ear, something that itself is attached to an alpha-Gal epitope, I've got him." That's the way the immune response works. It doesn't stick to the whole organism but basically finds some feature on it that's sticking out somewhere.

So I'm looking for something that's always going to be on an organism that I expect to be a problem soon. Staph has a neat little spot where it has to have a receptor for picking up iron when it's living in a human. It has one kind that picks up the haem group from our hemoglobin, and takes it inside the Staph cell. It's got to have that all the time. The reason is because Staph has to interact with a protein, created by the Staph, that goes out and gets the haem, brings it back, and docks with it. This feature of the Staph is always present on its surface, always conserved in the structure. If it messed with the structure—for example, mutating it fast—then it wouldn't fit with the docking protein, the thing couldn't get iron, couldn't grow inside you, and you wouldn't have a problem.

I look through the structural information that's already neen accumulated by the thousand people in the world who study *Staphylococcus* in a broad sense, and I say, "Well, here's a peptide, a 10-amino-acid peptide that looks like an unstructured kind of loop." It's wonderful that all that information is there, without having to go into a laboratory. Some people might find it boring, but I find it really exciting. I'm looking for the Achilles heel of any organism that needs to be taken off the street.

Once I find a possible target, colleagues can employ processes to discover relevant aptamer molecules to bind it. Our system uses aptamers, a relatively novel class of DNA/RNA binding molecules which were originally discovered in a systematic manner using the clever SELEX [Systematic Evolution of Ligands by EXponential enrichment] process, invented by Craig Tuerk 20 or so years ago, and are now being explored for a variety of applications through use of SELEX and other methods.

Aptamers will bind specifically to the target, and with high affinity—meaning that they attach to the given target but not to other targets and to that given target strongly. It's fairly complicated, but chemists have come a long way from turning lead into gold. They use a machine to make single-stranded DNA, not the double kind in a spiral but single strands. We want a known sequence of about 20 bases on the left end and a known sequence of 20 bases on the right end. Then in the middle we want about 30 bases, which are randomly selected from a bottle with all four bases in it. That means we have a potential for about 1018 different molecules in the same tube. That's more than the number of stars in the visible universe. Even one copy of each of them would not fit in a test tube, so we have to be content with about 1011, which is more like the number of stars in our own galaxy.

Some of them will bind to our target. They will be retained by a small column containing many copies of our target immobilized on a solid support. We wash the non-binders away, and using PCR, we make billions of copies of the binders. We can do this because we know what sequence they all have on their ends and we can make short primers to match these. Now we sequence a few of them completely, and with the sequence in hand, we can synthesize large amounts. All that's left to do is stick a synthetic alpha-Gal epitope on the aptamer and we have a drug. That's how it works, in theory.

When I first started working on this, about 10 years ago, the molecules we made were not at all stable in serum. There are lots of enzymes that destroy foreign DNA. It also seemed likely that the kidney would dispose of such a drug right away, once we started putting it in animals.

Following a suggestion from Jeeva Vivekananda, we have found several innovative ways to stabilize the aptamers in the circulation, and these are currently under further investigation. Anyway, that's how we explain their serum stability to ourselves and the fact that in our first *in-vivo* trials in rats, our drug that was designed to bind to the lethal factor, which is part of the anthrax toxin mechanism, saved the lives of the nervous rats who had been infected with a lethal dose of anthrax. It was an impressive experiment. We did it over and over again, and it definitely worked.

Now we're starting to work with organisms more likely to appear in a hospital, like Staph and influenza, and we have our sights on *Clostridium difficile*, *Pseudomonas aeruginosa*, *Acinetobacter baumanii,* and an alarming number of other bacteria that are resistant to antibiotics. We are also working on influenza, which has a convenient little feature called M2e. It's very promising, in my opinion, because the process for making the drug is prescribed completely.

A lot of different labs had to cooperate to make it. It's not something simple like PCR; when I invented that, I could do it all by myself. But in the case of anthrax, you've got to have a lab that's used to doing it or you'll end up killing yourself. You need an infectious-disease lab, and you need people who know how to raise and medically support small animals. It's a complicated process.

What I do personally is the research, which I can do from home because of the Internet, which pleases me immensely. I don't need to go to a library; I don't need to even talk to people face to face. I do travel to the labs. At Brooks Air Force Base in San Antonio, where we did the anthrax work, we are now working on a couple of strains of *E. coli* that are bothersome and dangerous.

Drugs that kill lots of different organisms breed many resistant strains of bacteria. They pass little things called plasmids freely around—plasmids contain instructions for making the resistance proteins. It's like somebody standing on the corner handing out leaflets, and not just to members of its own species. That's why resistance to organisms is spreading rapidly.

When Alexander Fleming first discovered penicillin, his boss, Almoth Wright, said bacteria would become resistant to it. It took longer than Wright thought, but it's happening. The narrowly directed drugs won't make bacteria resistant to them because they don't affect every other organism; they're not going to bother your *E. coli* or bother all the other organisms in your mouth and in your body. Once they're out of your body, they won't be effective at all. This is an important point. If you take penicillin, you excrete half of it. It goes down into the sewer in low, subclinical doses. It doesn't kill all the things in the sewer, but it definitely makes them start developing a resistance to penicillin. Most antibiotic resistance may not arise in our own bodies but elsewhere.

I've gone to the pharmaceutical companies with this concept, and they know it's a great idea. I expected them to buy in, but they didn't. It doesn't look like our drugs will make them $3 billion in the first year, which is their model. They like blockbuster drugs that people take on a daily basis. You spend the $200 million to get it approved, and then you have 10 years or so of an exclusive market on it. During that time, resistance to the drug might start happening, but you still have a proprietary product all the way through.

That's the economic model. It's not a good long-term strategy, because once you get resistance to one drug, you have resistance to that whole class of drugs, and there are most likely a limited number of classes. We're running out of antibiotics. But people will catch on. This is a good thing. We will produce a pipeline of antibiotic-type drugs. They're not really antibiotics, in the sense that they don't kill bacteria; they urge your immune system to do it. They say, "Eat me!"

16
Duck Sex and Aesthetic Evolution

Richard Prum

[September 3, 2014]

Richard Prum **is curator of ornithology and head curator of vertebrate zoology in the Peabody Museum of Natural History, Yale University.**

Over the last few years, I've realized that a large portion of the work I've been doing on bird color, on birdsong, on the evolution of display behavior, is really about one fundamental and important topic, and that's beauty—the role of beauty in nature and how it evolves. The question I'm asking myself a lot now is, What is beauty and how does it evolve? What are the consequences of beauty and its existence in nature?

There's a long history of people thinking about ornament in nature—those aspects of the body or the behavior of organisms that are attractive, that function in perception of other organisms. Usually we think about this in terms of sexual selection or mate choice, but there's a bunch of other contexts in which it can occur, like flowers attracting pollinators and fruits attracting frugivores. Or even the opposite—a rattlesnake or a poisonous butterfly scaring away predators. These are all aspects of the body that function not in the regular way but in perception.

For example, if we think about a plant and its parts, trying to explain why they are the way they are— If we examine the roots, for example, we could come up with a complete description of them and their physical function in the soil. They grab into the substrate, they absorb water and minerals, they help the plant anchor itself. They might even be interacting with fungi and bacteria in the soil. We have a theory for this, and that theory is natural selection. However, if we think about the flower, many parts of the flower—including its color, the shape of its petals, its fragrance—function through the perceptions of other animals. That is, the bee or the hummingbird comes along and regards the flower, asks itself, "Do I want to forage at that flower now?" and then either decides to do so or not.

To come up with a complete description of the function of a flower, we need a whole new kind of data. Not just a description of the physical world but something else inside, if you will—the mind of this other organism, this cognition. What I'm coming to conclude is that this is a watershed in evolutionary biology, and that there's a distinct process that occurs when we have evolution occurring through a cognitive or mental substrate—that is, when it's about attracting another individual.

I would refer to this area as aesthetic evolution, and the main topic in aesthetic evolution is the origin of beauty. Of course, these two words are not often involved in the sciences. In fact, science is a bit afraid of beauty, afraid of the aesthetic. This has to do with the fact that these terms refer to subjective experience. They refer to something that's sort of unknowable or immeasurable going on inside the cognitive capacities of other organisms. Of course we have difficulty understanding what's going on in somebody else's mind when they eat rhubarb pie, or smell a flower, or like or don't like a certain kind of music. Scientists are justifiably afraid of talking about subjective experience, and for the most part we've ceded this area to other fields, like the humanities. The way nature is—the nature of flowers, the nature of birdsong and bird plumages—implies that subjective experiences are fundamentally important in biology. The world looks the way it does, and is the way it is, because of their vital importance as sources of selection in organic diversity, so we need to structure evolutionary biology to recognize the aesthetic, recognize the subjective experience.

We'll never be able to nail it down exactly, as we do many scientific questions. We don't know what's going on in my brain or your brain with the experience of red, or what it's like to listen to a Mozart symphony and why some people might like certain things and others not, but in biology we have an interesting opportunity. For example, there are 10,000 or so species of birds in the world, and every single species of bird has a slightly different song and a different courtship display and a different way of attracting a mate and communicating socially. Those have all evolved as a result of subjective experience: "Do I like this mate or not?" Making a sensory perception, a cognitive evaluation, and then a choice. These elements—sensory perception, evaluation, and choice—give rise to this aesthetic evolutionary phenomenon.

Quantifying or describing the mental state of an individual, whether it's an individual bird listening to a birdsong or me, is very challenging. One of the things we can do in biology is to study the way in which subjective experiences evolve. We may not know what's happening in the brain of a particular bird species, but by studying the fact that it has diverged in its preferences from other warblers or other gulls or other woodpeckers indicates that we can study the evolution of subjective experience within comparative biology.

It's a little like the history of physics when they had a hard problem. You couldn't identify the velocity of the electron and its position all at the same time. What did they do? They had a hard problem, but they didn't say, "OK, we're going to shelve that problem for some other discipline to deal with." They created new tools—quantum mechanical mechanisms and quantum mechanical concepts—that would allow them to study the hard question of the position and velocity of the electron probabilistically.

In the same way, we can't know what's going on in the brain of any individual organism when it makes a subjective evaluation, but we can study how those preferences evolve. We can look at the history of diversification preferences. This is a new way to get into this area of the evolution of subjective experience in a way that people have been afraid of, and that's why evolutionary biology has a special role for understanding this aspect of nature, which I call the aesthetic.

A lot of scientists would probably still be allergic to the words "beauty" and "aesthetic" in science. We can progress without using those loaded words. But those loaded words, or what people think of as loaded words, are actually effective. They're communicating exactly what it is we want to get at, which is that powerful feeling of "Huh, I like that!" which motivates organisms across the board. Some people think, "Well, how can a bee have a subjective experience? How can they experience beauty? Clearly it's a simple system responding in a mechanistic way to stimuli, like flowers." If that were the case, if flowers were just specifically designed to push the buttons of bees, make them come and forage because they're so irresistible, then all flowers would converge on that same button, they'd be in the same form, they'd be pushing those bees' buttons and making them forage on that flower. But the fact is that flowers are diverse. They're all evolving to be memorably attractive, seductive. They're all appealing to that bee who says, "Huh, do I want to forage on that today?"

Of course to the bees some of those flowers are like nachos, cheap and irresistible, but others are special, like carrot cake, or whatever. You can go a long way looking for that thing and then enjoy it when you get it. That's what the bees are doing and that's why flowers are diverse, because of the existence and power of subjective experience as an agent in the evolution of organisms.

After Darwin described the mechanism of evolution by natural selection in the *Origin of Species*, he had a big problem, and that was the explanation of ornament in nature—those features of the body of animals and plants that function not in furthering the struggle for survival but in communication with other individuals, and often in the context of mating or ecological interactions. He got a lot of criticism for this, and so he was worried. He wrote to Asa Gray, a great American botanist, in 1861 and said, "The sight of a peacock's tail, whenever I look at it, makes me sick." Darwin was a sickly guy, and he took a lot of this very seriously. He took it to heart and worried and studied and a little over a decade later, in 1871, he wrote a second book, *The Descent of Man*, in which he described evolution by sexual selection.

Sexual selection was distinct from natural selection in that it had to do with differential reproductive success. Not survival up until the moment of mating but differential access to mates as a result of two possible mechanisms: One was male-male competition, or competition within the sex; the other was female choice, or mate choice of the one sex for members of the opposite sex. Darwin elaborated and predicted how male-male competition should give rise to armaments like antlers and large body size as in elephant seals, and that nature should give rise to ornaments like birdsong, beautiful bird plumage, and many other ornamental features. Darwin used explicitly aesthetic language to describe his theory. He described the mating preferences of birds as standards of beauty. He described female birds as having an aesthetic faculty. He described birds as the most aesthetic of all organisms—excepting, of course, man—and he was greatly criticized at the time.

In fact, his theory implied that female aesthetic judgments were a major force in evolution, and that was countered immediately by misogynistic responses describing female choice as "vicious feminine caprice." In those days, "vicious" meant "full of vice." In other words, it was even immoral, this theory. In particular, Darwin was criticized for proposing that there was some other theory that might explain evolution other than natural selection. His primary critic and his strongest and most persistent critic was the co-discoverer of natural selection, Alfred Russel Wallace. In the last decade or so of Darwin's life, Wallace and Darwin were duking it out over the meaning of sexual selection. Wallace was extremely negative about the prospect of beauty having any role in nature, and he criticized the theory in many ways, but he couldn't criticize it entirely. When he admitted that it could occur, he said, "Only under special conditions." And those special conditions would be when ornament was correlated with qualities that were demonstrably better in terms of natural selection—that is, longer life or better resources or better health. That is, when the trait had evolved some kind of meaning that the female would benefit from choosing.

Today we think of Wallace as the guy who killed sexual selection theory, but actually what Wallace did was describe for the first time the most popular model of sexual selection today, which is that ornament functions by providing a rich body of information about mate quality that mates need to know, and that mate choice is basically about improving conditions of the offspring.

As a scientist, I don't really care that much about this piece of history. Our job is to come up with and write theories that we have today. I'm interested in what Darwin and Wallace thought and their debates, but it's not critical to science. And yet Darwin still has such an important intellectual status in the world. This Darwin/Wallace debate is an interesting frame in which to think about the debate we should still be having.

I'm eager to revive the Darwin/Wallace debate in current biology—comparing Darwin's broader aesthetic perspective recognizing that sensory delight, attraction, subjective experience are really the agents of selection in these cases, and Wallace's honest advertisement of the quality-indication model in which the evolution of preference is controlled by a higher power—that higher power being adaptive natural selection. This is the debate that's happening in literature today and one I'm eager to inspire. In essence, what I hope to do is restore the Darwinian view, the legitimately Darwinian view.

In fact, the ideas of honest advertisement and quality indication were variantly anti-Darwinian back in the 1870s, and they still are. Modern neo-Darwinians are really neo-Wallaceans; they aren't Darwinian in the slightest, in the sense that they have laundered out of Darwin's legacy this history of a regard for aesthetics as an independent force in evolutionary biology, one potentially unhinged from natural selection.

There is this popular reductionist view of neuroaesthetics, which proposes, for example, that, through a combination of brain imagery and understanding neuro-function, we'll understand how the structure of the brain dictates what things will be attractive and not. This leads to a number of reductionist theories of aesthetics. For example, symmetry and indicators of symmetry are particularly important. What any review of art itself, in human arts or aesthetic features in organisms, will show is that there ain't no rules, and that rules are made to be broken, and that there is something irreducibly emergent about the way in which subjective experience evolves. And that has to do with what happens when you remove the controlling force of natural selection and allow subjective experience to be an independent player.

The theory on this was first developed by Ronald Fisher, who was an evolutionary biologist in the early 20th century and the inventor of statistics and the tea test. Most people and statisticians don't know that he was an evolutionary biologist. He came up with an idea: Imagine that some females like birds with red tails and others like birds with blue tails. Then you have males that have both blue and red tails. Well, not surprisingly, females who like red tails are going to find males who have red tails and females who like blue tails are going to find males who have blue tails. What happens is that as a result of selecting on male traits, mating preferences will become genetically correlated with the traits they prefer. That is, variation in desire and variation in the objects of desire will become correlated or enmeshed, entrained evolutionarily.

What that means is that when individuals, through the action of their preferences, select on traits, they're also indirectly selecting on their own preference. That means that preference is a self-organizing engine of evolution. That is, once you have popularity, then popularity itself can drive the evolution of ornament. Beauty and the desire for it and preferences co-evolve with one another; they are changing one another. The peacock's tail as it evolves is transforming the female's brain and her capacity to understand what beauty is, and her preferences are also transforming the tail; they evolve along an entrained path together.

Different species are all evolving in different directions, and that's why nature looks the way it does.

The sexual selection mechanism I'm interested in—or that I'm a big fan of, really—goes from Darwin to Fisher into, more recently, mathematical genetic models by Russ Lande and Mark Kirkpatrick and the intellectual lineage of that group. Of course, the opposite have gone from Wallace to a reinvention by [Amotz] Zahavi to lots of modern notions about how ornament should evolve to be honest advertisements. We have this arbitrary mechanism where natural selection is not really involved, and we have an honest mechanism where natural selection is the controlling force. These theories have been around for a while, and of course there's been lots of conflict.

My take has been to observe that the adaptationists now rule by essentially rejecting falsification. It's almost a faith-based enterprise, in the sense that what people do is they go to nature and they examine a trait—whether it's a patch of plumage or a color of a feather or a birdsong—to try to show that it's somehow correlated with some indicator of quality or some measure of direct benefits or good genes. And when they fail to do so, they conclude, "Oh, we're still right. We just haven't worked hard enough to show how it could be true." And when they find that it *is* true, they say, "Ah-ha, our theory is confirmed." As a result, what we have in the literature is like a weird bonsai tree. It's composed only of the examples that fit the theory. All of those failures to conform to the adaptive theory are evidence completely consistent with the arbitrary model.

They've protected themselves from this, and of course this goes back to a quote by Alan Grafen, who said, "To believe in the Fisher/Lande process as an explanation of sexual selection without abundant proof is methodologically wicked." Not many ideas even in evolutionary biology have been described as wicked, but arbitrary sexual selection is one of them. There's a lot riding on this, and my gambit recently is to propose that the Fisher model—the arbitrary model, the Lande/Kirkpatrick model—is essentially the null hypothesis. It's the prediction of the consequence of genetic variation in traits and preferences in the absence of natural selection on preferences. And so it's what we should expect a large part of the time.

In essence, the null model or the null mechanism is a tough sell, because basically this is the "shit happens" idea. Well, the latest version of this, of course, is that "beauty happens," and that's my new mantra. Right? But in essence it's a hard sell because a lot of people in science, and particularly in evolutionary biology, got into the field because of the buzz they got by explaining things in terms of adaptation. They're eager to confirm that model, and they see the other model as basically intellectually unsatisfying—or potentially wicked, even.

It's a tough sell, but like the neutralist selectionist debate that went on in genetic evolution or community assembly and community ecology, you can't do this kind of science in the absence of a null hypothesis. There is a perfectly legitimate and worthwhile null mechanism for the evolution of the diversity of singles and aesthetic experiences, and that is the arbitrary Lande/Kirkpatrick model.

I'm proposing this. I hear a lot of silence. People will read the paper and say, "Well, I liked what you said up to here, but then I had this problem." I can resolve those problems quickly, but yet there isn't this big movement to adopt null models in sexual selection. What will have to happen? A lot of it is about differential recruitment—bringing new people into the field who are fascinated by the prospect of studying aesthetic evolution, and that's why I'm here talking to you today.

The adaptationist approach views ornament as an embodied piece of information, quality information. The arbitrary model is the one in which traits and preferences co-evolve with one another purely in an aesthetic fashion and without anything other than the benefits of popularity. The adaptationist position is a lot like the efficient-market hypothesis in economics. That is, that the value of a commodity—how worthwhile this mate might be or the value of your house—is explicitly measureable and will always approach a real value because all the players are honest and rational.

The aesthetic approach or the arbitrary model is a lot like the irrationally exuberant market bubbles. And going all the way up to the moment of the crash of the housing market and the consequence economic disaster across the globe in 2007 and 2008, you had efficient-market theorists saying that bubbles were impossible, they couldn't exist, and even describing them was a silly exercise. These guys are like the ardent adaptationists who would describe the Fisher/Lande hypothesis as methodologically wicked.

In economics, a lot of people who got into these views were the kind of people who read too much Ayn Rand as children, and they go out into the world to find the environment that supports the kinds of thoughts they have. I maintain that a lot of people in evolutionary biology have come into evolutionary biology because they were attracted to the concept of adaptation. They were influenced to go into the field so they could have this buzz of explaining the complex in terms of a lawlike overriding idea, and since aesthetics can co-evolve in lots of different directions in different species, it evades that, and that's why Wallace got so exercised about Darwin. That's why Alan Grafen described the Fisher/Lande mechanism as methodologically wicked. It is a real existential threat to the global lawlike power of natural selection to explain biodiversity.

I'm an evolutionary biologist. I've done evolutionary biology for my entire career almost exclusively, and yet I have been realizing in recent years that for the most part I find adaptation to be kind of boring. I know it's ubiquitous and I know it's important and all, yet as an idea, as an intellectual concept, for me it's mostly over.

What's really interesting is that the contingency of history and biodiversity give rise to all sorts of things that are much more complicated, much more quirky, and much more fascinating than the lawlike property of adaptation. This has affected lots of my work, where I've been doing phylogenetics, evolution, and development; the physics of color production; and all these areas where contingency and history have a controlling force. What do you call all that together? I don't really know. It's an evolutionary structuralism, where the contingency of history is an important principle in how evolution proceeds.

One of the important consequences of an aesthetic view in evolutionary biology has to do with the 20th-century history of evolution and that period when Wallace was successful in redefining all sexual selection as merely a form of natural selection, so there was no theory of mate choice, no aesthetic theory available. This is basically the period between 1880 and 1970.

One of the interesting features during this period—when evolutionary biology had no concept of mate choice and all mating was under the control of natural selection—is that that's also the period in which evolutionary biology was dominated by eugenic theory. What people don't like to think about is that essentially every evolutionary biologist in the early 20th century was either an ardent eugenicist or a happy fellow traveler. This is exactly the period in which many of the concepts we still use in evolutionary biology were codified and defined. The core mathematical theory of population genetics was created. Those influences don't just go away when we say, "Oh, we're no longer eugenicists."

If you stick with the hard-line adaptationist view of sexual selection, in which mate choice is always for those features of your mate that indicate quality, then essentially all mate choice is about natural selection. It's all about getting ahead. In contrast, if you have an aesthetic theory, where sometimes things, merely because they're popular, evolve to be more present, you have essentially an opportunity for decadence to evolve, for not just costly honest advertisements but things that are genuinely costly.

By adopting a null hypothesis, a null model of mate choice, or evolution by mate choice, in which arbitrary, aesthetic traits can evolve, you permanently inoculate evolutionary biology from this eugenic past. That's critically important, because it's not gone away just because we'd like it to. We have to build a science that prevents us from doing eugenic science again, and that's when you think about genomics and you think about evolutionary psychology and the lack of a null model or the possibility of an aesthetic null in any of that research program; it's a real and present problem. And that's something that could be transformative for evolutionary biology.

Eugenics was the science of human racial superiority—that certain races had evolved, as a consequence of natural selection, to be superior to others. It led to a concern both for genetic superiority—"eugenic" means, you know, "well born"—and true genes (and that's, essentially, good genes) as one of the honest advertisement theories. Almost the same terminology. Eugenics was also concerned with class and money and the environment.

These are what we now refer to as direct benefits. Both of the concerns of eugenics are still actively involved in adaptationist theories about mate choice.

Right now, the field of sexual selection and the evolution of mate choice is dominated by the adaptationist school—the Wallacean position that all traits and preferences evolve because they're correlated with features that are legitimately explicitly better. The arbitrary position has not got a lot of play in the last few years, so I'm trying to create a new way of looking at the field, to destabilize it by proposing that null models ought to be used and force a new scientific standard on it so that this confirmationist science—the idea that the only things we get published are the things everybody is comfortable with already—changes a little bit.

This is like the neutralist selectionist debate in population genetics, which occurred back in the '60s and '70s, and that's the kind of debate I'd like to start; the history there intellectually demonstrates that you can't do evolutionary science without a null model, and that means there's only one way this will go, and that is legitimately accepting a broader aesthetic definition of how mate choice could work, restricted in some circumstances to the adaptationist position, and that's where I'd like to see it head. It will happen only through differential recruitment—bringing new people into the field who are attracted to evolutionary biology specifically because it can do this thing that evolutionary biology cannot do right now.

One of the interesting consequences of the adaptationist view of sexual selection—the Wallacean view—is that we don't really need an account of why females prefer what they do; we don't need to focus our attention on the female as an evolutionary agent. The reason is because we have a larger, broader, powerful theory about adaptation that describes what the female is doing. We don't actually have to construct a theory where we recognize aesthetic agency—the capacity of females to influence the evolution of their species.

If you eliminate natural selection, or admit that sometimes it's present and sometimes it's not, then we have to ask, What are females doing? Why do they choose the preferences they do? And this has given rise in my own work to a series of fascinating research programs in the area of sexual conflict—that is, what happens when mate choice and mate competition conflict with one another? A perfect example of this is the case of waterfowl, or ducks; in ducks the males do these elaborate displays. They have the bright green head—"quack, quack, quack," and make way for ducklings—all the little movements they do. The females choose on the basis of those displays, and thus different duck species have different plumages and different colors to their body.

Meanwhile, there's another force going on. It turns out that there's a lot of male-male competition; and, indeed, this goes back to some deep reproductive biology of the ducks, which is that ducks are one of the few birds that still have a penis. It's a weird structure. It has an explosive erection, its erection mechanisms are lymphatic instead of vascular, it's stored outside-in inside the cloaca and comes flying out, and they can get very lengthy—up to 40 centimeters, which is over a foot long, on a duck that is itself not even a foot long. It's an extraordinary piece of biology. What's going on in these ducks?

In lots of ducks, there's forced copulation. It's the equivalent of rape in ducks. In species where there is a lot of forced copulation, females have evolved or co-evolved complex vaginal morphologies that frustrate the intromission—frustrate entry of the penis during forced copulation. For example, the penis of ducks is counterclockwise coiled and often has ridges or even teethlike structures on the outside. In this case, in these species, the female has evolved a vagina that has dead end cul-de-sacs, so that if the penis goes down the wrong direction, it'll get bottled up and isn't perceived to be closer to the oviduct, or further up the oviduct and closer to ova. And then above the cul-de-sacs, the duck vagina has clockwise coils, so there's literally anti-screw devices that prevent intromission during forced copulation.

What's happened here is that females have evolved the ability to prevent fertilization due to forced copulation most of the time. This is an indication of an evolution of a female advantage through sexual conflict. How does this happen, and what does this have to do with beauty? Well, the way in which it happens is, imagine that a female obtains the mate she desires. He's got the bright green head, he's got the "quack, quack, quack" that she loves. Her male offspring are going to inherit the genes for those attractive traits she likes, and her male offspring will therefore benefit by being preferred by other females of the species who evolved the same preferences.

If she's forcibly fertilized by rape, then her offspring will inherit either a random trait because it didn't pass the test of her preference, or they will inherit traits that have been specifically rejected by other females, and therefore her male offspring will not be as sexually attractive to other females, and that's a genetic cost to her. That's an indirect genetic cost to her fitness. As a consequence, she'll suffer a cost to sexual coercion. Those individuals that have vaginal morphologies allowing them to achieve what they desire will benefit, because other females will reward them by preferring their offspring. This is about how the evolved normativity—the co-evolved concept, if you will—of what is sexy, what females prefer, provides leverage that females can use to evolve, to expand their sexual autonomy, to expand their control, their agency in the face of sexual violence.

In the case of the ducks, one of the problems with this mechanism is that it's purely defensive, so the female evolves a more complicated vagina and the male evolves a bigger penis. She gets even more counterclockwise or clockwise coils, and he'll evolve a penis that has thorny tooth-like structures on it. It's an arms race. And that's not good. That's a lot of wasted investment.

There is, however, an alternative, and that's when female choice can act aesthetically to remodel or aesthetically transform male behavior in a fundamental way. A great example is the bowerbird. The bowerbirds are frugivorous—fruit-eating—tropical birds from Australia and New Guinea and nearby islands. What happens is the male builds a construction—a bower. Often it consists of two walls of sticks with a passageway in the middle, and he gathers all these materials, sometimes bones or berries or fruits or flowers that are brilliantly colored or sometimes a pile of snail shells, and the female visits the bower and chooses her mate based on the bower architecture and on the materials he provides.

In this case with the bowerbirds, the male builds a seduction theater at the behest of the female. It's evolving because females prefer it. They want this aesthetic architecture in order to advance their mate choices. What's interesting, of course, is that these structures are aesthetic, but they have this special property, which is that if the female is sitting inside the walls of the bower and the male displays in the front and is showing off his cool stuff, before he can copulate with her he has to go back around the back of the bower, and that gives her a chance to pop out the front. She is essentially protected from date rape.

She can explore and see him at a close distance, and all of his stuff, and yet she's protected from being jumped or sexually harassed by the male. And why? Because she has preferred those structures that provide her with a safe refuge in which to fulfill her aesthetic desires. In this case, we have in bowerbirds the evolution of aesthetic architecture which prevents forced copulation in the females, and that's another way in which females can use the evolved normative concept of what is beautiful to advance their autonomy.

Through thinking about duck sex and aesthetic evolution, I started to develop a scientific concept of sexual autonomy—the way in which a co-evolved concept of what's attractive provides leverage for the advancement of the freedom of choice, the freedom of mate choice. I've worked on birds all my career, with some little digressions on butterflies and beetles, but basically I'm an ornithologist. That has led me to entertain the possibility of aesthetic evolution in response to sexual conflict in human evolution. One of the things that's notable about humans is the transformation of male violence. We're a species where 98 or 99 percent of the violence is still the result of male behavior, and yet we're notably less violent than our closest relatives, like chimpanzees and gorillas. One of the particular ways in which we're less violent is the moderation of sexual conflict in humans.

To understand where we got where we are, let's imagine your typical Old World monkey or gorilla or chimpanzee. The situation for females is pretty grim. There's some male that's in political and social control of your group, or of you, and he controls most of your sexual life. What happens then is that occasionally there's social unrest, that male will be deposed, and a new male will come in. One of the first things these new males do is go out and kill all the babies, kill all of the dependent young. Why? Because lactation, or breastfeeding, prevents ovulation, prevents reproduction. By killing all the offspring of the previous male, he is advancing his reproductive opportunity. All the females will go into estrus, and he will have even sooner an opportunity to advance his own fitness. It's an incredibly selfish male behavior that evolves by male-male competition, and it has a huge negative impact on females.

Back in the '80s, Sarah Blaffer Hrdy and others established that one of the responses females have to this sexual conflict is to mate multiply with other males in hopes of buying an insurance policy, so that should that guy become the dominant male, he'll be less likely to kill her child, because he might be its father. But of course like ducks, this just gives rise to an arms race. If the female starts having sex with other males, then of course the dominant male is going to be much more likely to respond with force to reinforce his social control. Females are mating multiply, but it's not because they have desire that they're fulfilling, they're just trying to make the best of a bad situation. They're trying to prevent the killing of their offspring.

What's changed about people? Well, your average gorilla or chimpanzee is almost an infanticidal maniac waiting for his moment. Humans are pretty bad. We've enslaved each other, we have wars where we wipe each other out, but one of the crimes you don't read about in the newspaper is males killing children for their own reproductive advantage.

I'm interested in the possibility that aesthetic mate-choice in humans—female choice—could have played a critical role in the remodeling of male-male competition, essentially by establishing that those features of males that are associated directly with violent competition are unsexy—or, more positively, that those features associated with advancing female autonomy evolved to be a new form of sexy. That's the kind of dynamic interaction you get between sexual conflict and aesthetic mate choice that we see in birds like bowerbirds and lekking birds and throughout the bird world.

What would these traits be? Well, one of the interesting things is that even though human beings evolved to be much larger than their chimplike ancestors in body size, they actually have gotten less different in size. [Human] males and females are more similar in size than are [male and female] chimpanzees. This is exactly against the laws of allometry, which indicate that, as you get bigger, any differences between the sexes should get broader. That means there's been active selection to reduce the difference in body size between males and females, and that's likely to have evolved through female mate choice.

Another example is the human smile. All you've got to do is look at our smile. We have canine teeth that are sexually monomorphic. In all of our Old World primate relatives, including our closest relatives, the gorillas and chimpanzees, the males have deadly weapons in their mouths, which human males lack. The question is, under what conditions are males going to give up their weapons? The answer is difficult. You've got to get them below the belt. It's interesting that of course the Greeks thought of this first, in the play *Lysistrata* (performed in Athens in 411 BC), in which Lysistrata organizes the women of Greece to stage a sex strike. They won't have any sex until the men call off the war against Peloponnesia. Eventually after lots of comedy the males relent and the war is over and everybody goes back to sleeping together.

What's perceived, interestingly and appropriately, is that females, organized together, can transform male-male social relationships through mate choice. That is, the importance of bromance before romance. There's something about male cooperation which is particularly attractive to women, and that has had a transforming effect. This is an important issue. Everything we know that's distinctive about human biology is predicated upon the lengthening of the childhood, child dependency, and parental investment, whether it's brain size, the development of a language capacity, culture and the capacity to learn culture, material culture, technology. All of this required having longer growing-up times and a longer time to get smarter and smarter individuals.

Solving the infanticide problem was a big deal. It would be very difficult to evolve to invest more in every offspring if potentially a third or a quarter of all offspring are murdered as a result of social violence. And so perhaps one of the key features in the evolution of humanity is the solving of sexual conflict and the infanticide problem, and that's why aesthetic evolution and its interaction with sexual conflict has a fascinating role to play in understanding human nature and the evolution of human biology.

I started bird watching as a child about ten years old, when I got my first pair of glasses and suddenly the earth came into focus around me and within six months I was a bird watcher. I started with the listening. I was living in a small town in southern Vermont at that point, and I tramped over hill and dale and all around to see as many birds as I could, and I got hooked on biodiversity. One of the things, as a kid, when you start learning birds and learning birdsongs, you're establishing mental circuits—a way of your brain working in the world that captures this kind of knowledge in an efficient way. It becomes how you think. When I got to college, I knew I was going to be involved in ornithology, and I actually imagined myself becoming a park ranger or running a refuge. I thought that's what it [ornithology] was. I had a good education, but I didn't know what science was, as a life.

Then I discovered that evolutionary biology was the field of science that was about what I was interested in, which was biodiversity and the origin of all the different birds I had been learning. I soon got influenced and interested in phylogenetics, the reconstruction of bird phylogeny, which was the revolutionary event that was happening right at that time. Ultimately I tried to combine my experience of bird watching with my interest in phylogenetics, and I ended up studying the evolution of courtship display in a family of South American birds called manakins, spending a lot of time in the jungle describing the courtship dances of birds that prior to that time were only known from museum drawers, that were not really known in life, and that was delightful.

Part of that work was listening and learning birdsong, and even from a young age I was always able to do that. Starting in grad school, I had a sudden idiopathic hearing loss, probably viral, in my right ear. It took out everything above 1,500 hertz. That's about the middle high range on the righthand side of a piano. Overnight. About five or ten years later, I started to develop what's called Meniere's disease in the opposite ear, which is a problem of control of the endolymph, the fluid inside the ear, and ultimately lost all that hearing. In the middle of my thirties, having made a career out of studying birds in the field and studying their behavior, I suddenly couldn't hear them any longer. Right now I'm pretty much ornithologically deaf. I can hear a crow, I can hear the bottom half of a robin, but most of the birdsongs of the world I can no longer hear. I suppose I could work on penguins, but that's not the kind of fieldwork I used to do, which was working in the jungle acoustically. In the middle of my career I had to develop a new connection to my life's work, and that was a big challenge, but that led to a fantastic series of research programs on feathers, on the evolution of feathers, on color, which I can still, fortunately, see really well, and a whole new kind of research emerged out of that.

At the core of aesthetic evolution is the idea that organisms are aesthetic agents in their own evolution. In other words, birds are beautiful because they're beautiful to themselves. And that scientific conclusion has the power to transform our relationship to nature as people who can walk around in nature and regard flowers and listen to birdsong and watch birds and appreciate them in a new way. Aesthetic evolution as a scientific concept has the power to transform how we experience nature itself, and I know that my own birdwatching has been transformed by this. When I'm looking at an indigo bunting, which is a beautiful blue bird, or a scarlet tanager, which is brilliantly red with black ring patches and black tail, then imagining how they came to be through this co-evolutionary dance between traits of the male and the preferences of the female transform what that's like.

When you're listening to a wood thrush's complex fluty song and realizing the aesthetic process that's given rise to that, it has a transformative effect. I'm hoping that this view of nature gets out to the public and changes the way in which we look at nature, and gets people out more often to learn birdsong themselves and pick up that field guide, identify those birds at the feeder or that are migrating through in the spring. And even though my own brain includes probably hundreds if not thousands of neurons dedicated to learning and knowing birdsongs that I can no longer hear, this exercise of going out into nature and observing it as a human being, by understanding the science and the aesthetic lives of the organisms themselves, is a really special experience. I hope this work will encourage people to do that.

17
Toxo

Robert Sapolsky

[December 2, 2009]

Robert Sapolsky is Gunn Professor and professor of neurology and of neurosurgery, Stanford School of Medicine.

In the endless struggle that neurobiologists undergo—in terms of free will, determinism—my feeling has always been that there's not a whole lot of free will out there, and if there is, it's in the least interesting places and getting more sparse all the time. But there's a whole new realm of neuroscience I've been thinking about, which I'm starting to do research on, that throws in an element of things going on below the surface. It's got to do with the bizarre world of parasites manipulating our behavior. It turns out that this is not all that surprising. There are all sorts of parasites out there that get into some organism, and what they need to do is parasitize the organism and increase the likelihood that they will be fruitful and multiply—and in some cases they can manipulate the behavior of the host.

Some of these are pretty astounding. There's a barnacle that rides on the back of a male crab and can inject estrogenic hormones into it, at which point the male's behavior becomes feminized. The male crab digs a hole in the sand for his eggs, although he has no eggs. But the barnacle sure does and has got this guy to build a nest for him. There are cases where wasps parasitize caterpillars and get them to defend the wasps' nests for them. These are extraordinary examples.

The parasite my lab is beginning to focus on is one in the world of mammals. It's this protozoan called *Toxoplasma*. If you're pregnant, if you're around anyone who's pregnant, you know you immediately get skittish about cat feces, cat bedding, cat everything, because they could carry Toxo and you do not want *Toxoplasma* to get into a fetal nervous system. It's a disaster.

The normal life cycle for Toxo is one of those amazing bits of natural history. Toxo can reproduce sexually only in the gut of a cat. It comes out in the cat feces. The feces get eaten by rodents. So Toxo's evolutionary challenge at that point is to figure out how to get rodents inside cats' stomachs. Now, it could have done this in unsubtle ways, such as crippling the rodent or some such thing. Toxo instead has developed an ability to alter innate behavior in rodents.

If you take a lab rat who is 5,000 generations into being a lab rat, whose ancestor actually ran around in the real world, and you put some cat urine in one corner of the rat's cage, the rat will move to the other side. Completely innate, hard-wired reaction to the smell of cats, the cat pheromones. But take a Toxo-infected rodent, and it will no longer be afraid of the smell of cats. In fact, it will become attracted to the smell of cats. The most amazing damn thing you'll ever see! Toxo knows how to make cat urine smell attractive to rats—and rats go and check it out. That rat is now much more likely to wind up in a cat's stomach. Toxo's circle of life is completed.

This experiment was reported by a group in the U.K. about half a dozen years ago. Not a whole lot was known about what Toxo was doing in the brain, so ever since, part of my lab has been trying to figure out the neurobiological aspects. The first thing is that it's for real: The rodents—rats, mice—really do become attracted to cat urine when they've been infected with

Toxo. And you might say, "OK, well, this is a rodent doing all sorts of screwy stuff because it's got this parasite turning its brain into Swiss cheese; it's just nonspecific behavioral chaos." But no, these are normal animals. Their olfaction is normal, their social behavior is normal, their learning and memory is normal. All of that. It's not a generically screwy animal.

You say, "OK, well, it's not that, but Toxo seems to know how to destroy fear and anxiety circuits." But it's not that, either. These rats are still innately afraid of bright lights. They're nocturnal animals. They're afraid of big, open spaces. You can condition them to be afraid of novel things. The system works perfectly well there. Somehow, Toxo can laser out this one fear pathway, this aversion to predator odors.

We started looking at this. The first thing we did was introduce Toxo into a rat and it took about six weeks for the Toxo to migrate from its gut up into its nervous system. At that point, we looked to see where it had gone in the brain. It formed cysts—latent, encapsulated cysts—and it wound up all over the brain.

That was deeply disappointing. But when we looked at how much wound up in different areas in the brain, it turned out that Toxo preferentially knows how to home in on the part of the brain that's all about fear and anxiety—a region called the amygdala. The amygdala is where you do your fear conditioning. The amygdala is what's hyperactive in people with post-traumatic stress disorder. The amygdala is all about pathways of predator aversion, and Toxo knows how to get in there.

Next, we saw that Toxo would take the dendrites—the branch and cables with which neurons connect to one another—in the amygdala and shrivel them up. The Toxo was disconnecting circuits; you wound up with fewer cells there. This is a parasite that's unwiring in the part of the brain critical for fear and anxiety—

that's really interesting! But it doesn't tell us a thing about why *only* the predator aversion has been knocked out whereas fear of bright lights, et cetera, is still in there. Toxo knows how to find a particular circuitry!

So what's going on here? What's it doing? Because it's not just destroying this particular fear-aversive response, it's also creating something new—an attraction to cat urine. And here is where this gets utterly bizarre. You look at circuitry in the brain, and there's a reasonably well-characterized circuit that activates neurons which become metabolically active circuits, where they're talking to one another—a reasonably well-understood process involved in predator aversion. It involves neurons in the amygdala, the hypothalamus, and some other brain regions getting excited. This is a very well-characterized circuit.

Meanwhile, there's also a well-characterized circuit having to do with sexual attraction. And part of that circuit courses through the amygdala—which is pretty interesting in and of itself—and then goes to different areas of the brain. When you look at normal rats and expose them to cat urine, cat pheromones, they show a stress response, exactly as you would expect. Their stress-hormone levels go up, and they activate this classical fear circuitry in the brain. Now you take Toxo-infected rats right around the time when they start liking the smell of cat urine, you expose them to cat pheromones, and you don't see the stress-hormone release. What you see is that the fear circuit doesn't activate normally, and *instead* the sexual arousal circuit activates. In other words, Toxo knows how to hijack the sexual-reward pathway. You get males infected with Toxo and expose them to a lot of the cat pheromones, and their testes get bigger. Somehow, this parasite knows how to make cat urine smell sexually arousing to rodents, and they go and check it out. Totally amazing! So on a certain level, that explains everything: "Ah-ha! It takes over sexual-arousal circuitry."

At this point, we don't know what the basis is of the attraction in the females. It's something we're working on.

Some nice work has been done by a group at Leeds in the U.K., which is looking at the Toxo genome, and we're picking up on this collaboratively. Toxo is a protozoan parasite. Toxo and mammals had a common ancestor, and the last ancestor they shared was, god knows, billions of years ago. You look in the Toxo genome, and it's got two versions of the gene called tyrosine hydroxylase. And if you were a neurochemistry type, you would be leaping up in shock and excitement at this point.

Tyrosine hydroxylase is the critical enzyme for making dopamine, the neurotransmitter in the brain that's all about reward and anticipation of reward. Cocaine works on the dopamine system; all sorts of other euphoriants do. Dopamine is about pleasure, attraction, and anticipation. And the Toxo genome has the mammalian gene for making the stuff. It's got a little tail on the gene that specifies that when this is turned into the actual enzyme, it gets secreted out of the Toxo and into neurons. This parasite doesn't need to learn how to make neurons act as if they are pleasurably anticipatory; it takes over the brain chemistry of it all on its own.

Again, that issue of specificity comes up. Look at parasites closely related to Toxo: Do they have this gene? Absolutely not. Now look at the Toxo genome and look at genes related to other brain messengers—serotonin, acetylcholine, norepinephrine, and so on. Go through every single gene you can think of. Zero. Toxo doesn't have them. Toxo's got this one gene which allows it to plug into the whole world of mammalian reward systems.

At this point, you'll ask, "Well, what about other species?

What does Toxo do to humans?" And there's some interesting stuff there, reminiscent of what goes on in rodents. Clinical dogma is, you get a Toxo infection and if you're pregnant, it gets into the fetal nervous system, a huge disaster. Otherwise, if you get a Toxo infection, it has phases of inflammation but eventually goes into a latent, asymptomatic stage, which is when these cysts form in the brain. Which is, in a rat, the stage at which it stops being asymptomatic and the strange behavior starts occurring. Interestingly, that's when the parasite starts making the tyrosine hydroxylase.

So, what about humans? A small literature is coming out now reporting neuropsychological testing on men who are Toxo-infected, showing that they get a little impulsive. Women less so, and this may have some parallels, perhaps, with the testosterone aspect of the story we're seeing. Here is the truly astonishing thing: Two different groups independently have reported that people who are Toxo-infected have three to four times the likelihood of being killed in car accidents involving reckless speeding.

In other words, you take a Toxo-infected rat and it does some dumb-ass thing that it should be innately skittish about, like going right up to cat smells. Maybe you take a Toxo-infected human and they start having a proclivity toward doing dumb-ass things that we should be innately averse to, like having your body hurdle through space at high G-forces. Maybe this is the same neurobiology. This is not to say that Toxo has evolved the need to get humans into cat stomachs. It's just sheer convergence. It's the same nuts-and-bolts neurobiology in us and in a rodent, and does the same thing.

On a certain level, this is a protozoan parasite that knows more about the neurobiology of anxiety and fear than 25,000 neuroscientists standing on each others' shoulders, and this is not a rare pattern. Look at the rabies virus; rabies knows more about aggression than we neuroscientists do. It knows how to make you rabid. It knows how to make you want to bite someone, and that saliva of yours contains rabies virus particles, [ready to be] passed on to another person.

The Toxo story is, for me, new terrain—totally cool, interesting stuff, just in terms of this individual problem. And maybe it's got something to do with treatments for phobias down the line. But no doubt it's also the tip of the iceberg of other parasitic stuff going on out there. Even in the larger sense, god knows what other unseen realms of biology make our behavior far less autonomous than lots of folks like to think.

With regard to parasitic infections like Toxo in humans, there's a higher prevalence in the tropics, where typically more than half the people are infected. Rates are lower in more temperate zones, for reasons I don't understand and do not choose to speculate on. In much of the developing world, it's from bare feet absorbing the Toxo through soil where cats may have been. It's from food that may not have been washed sufficiently, and absorption through hands.

A few years ago, I sat down with a couple of the Toxo docs in our hospital who do the Toxo testing in Ob/Gyn clinics. They hadn't heard about this behavioral story, and I'm going on about how cool and unexpected it is. And suddenly one of them jumps up, flooded with 40-year-old memories, and says, "I just remembered, back when I was a resident, I was doing a surgical-transplant rotation. And there was an older surgeon who said, 'If you ever get organs from a motorcycle-accident death, check the organs for Toxo. I don't know why, but you find a lot of Toxo.'"

What's the bottom line? Well, it depends; if you want to overcome some of your inhibitions, Toxo might be a good thing to have in your system. Not surprisingly, ever since we started studying Toxo in my lab, every lab meeting we sit around speculating about which people in the lab are Toxo-infected and [whether] that might have something to do with someone's level of recklessness. Who knows? It's interesting stuff, though.

You want to know something terrifying? Here's something terrifying and not surprising. The U.S. military knows about Toxo and its effect on behavior. They're interested in Toxo. They're officially intrigued. And I would think they would be intrigued, studying a parasite that makes mammals perhaps do things that everything in their fiber normally tells them not to do because it's dangerous. But suddenly, with this parasite on board, the mammal is a little more likely to go and do it. Who knows? But they're aware of Toxo.

There are two groups that collaborate in Toxo research. One is Joanne Webster's, who was at Oxford when she first saw this behavioral phenomenon, and she's now at Imperial College London. The other is Glenn McConkey's, at the University of Leeds. She's more of a behaviorist; he's more of an enzyme biochemist guy. We're doing the neurobiology end of it. We're all talking lots.

There's a long-standing literature showing a statistical link between Toxo infection and schizophrenia. It's not a big link, but it's there. Schizophrenics have higher than expected rates of having been infected with Toxo, and that's not particularly the case for related parasites. There are links between schizophrenia and mothers who had house cats during pregnancy. There's a whole literature on that. And whenever Toxo is picked up in the media and this schizophrenia angle is brought in, the irresistible angle is the generically crazy cat lady—you know, living in the apartment with 43 cats and their detritus.So where does this fit in?

Two really interesting things: Back to dopamine and the tyrosine hydroxylase gene that Toxo somehow ripped off from mammals and which allows it to make more dopamine. Dopamine levels are too high in schizophrenia. That's the leading suggestion of what schizophrenia is about neurochemically. The brains of Toxo-infected rodents have elevated levels of dopamine. Final deal is—and this came from Webster's group—you take a rat who's been Toxo-infected and is now at the stage where it would find cat urine attractive, and you give it drugs that block dopamine receptors, the drugs that are used to treat schizophrenics, and it stops being attracted to the cat urine. So there's some schizophrenia connection here.

18
The Adjacent Possible

Stuart Kauffman

[November 9, 2003]

Stuart Kauffman is a theoretical biologist and founding director, Institute for Biocomplexity and Informatics, University of Calgary

In his famous book *What is Life*?, Erwin Schrödinger asks, "What is the source of the order in biology?" He arrives at the idea that it depends on quantum mechanics and a microcode carried in some sort of aperiodic crystal—which turned out to be DNA and RNA—so he is brilliantly right. But if you ask if he got to the essence of what makes something alive, it's clear that he didn't. Although today we know bits and pieces about the machinery of cells, we don't know what makes them living things. However, it is possible that I've stumbled upon a definition of what it means for something to be alive.

For the better part of a year and a half, I've been keeping a notebook about what I call autonomous agents. An autonomous agent is something that can act on its own behalf in an environment. Indeed, all free-living organisms are autonomous agents. Normally, when we think about a bacterium swimming upstream in a glucose gradient, we say that the bacterium is going to get food. That is to say, we talk about the bacterium teleologically, as if it were acting on its own behalf. It is stunning that the universe has brought about things that can act in this way. How in the world has that happened?

As I thought about this, I noted that the bacterium is just a physical system; it's just a bunch of molecules that hang together and do things to one another. So, I wondered, what characteristics are necessary for a physical system to be an autonomous agent? After thinking about this for a number of months, I came up with a tentative definition. My definition is that an autonomous agent is something that can both reproduce itself and do at least one thermodynamic work cycle. It turns out that this is true of all free-living cells, except for weird special cases. They all do work cycles, just like the bacterium spinning its flagellum as it swims up the glucose gradient. The cells in your body are busy doing work cycles all the time.

Once I had this definition, my next step was to create and write about a hypothetical chemical autonomous agent. It turns out to be an open thermodynamic chemical system that can reproduce itself, and, in doing so, performs a thermodynamic work cycle. I had to learn about work cycles, but it's just a new class of chemical-reaction networks that nobody's ever looked at before. People have made self-reproducing molecular systems and molecular motors, but nobody's ever put the two together into a single system capable of both reproduction and doing a work cycle.

Imagine that inside the cell are two kinds of molecules—A and B—that can undergo three different reactions. A and B can make C and D, they can make E, or they can make F and G. There are three different reaction pathways, each of which has potential barriers along the reaction coordinate. Once the cells make the membrane, A and B can partition into the membrane, changing their rotational, vibrational, and translational motion. That, in turn, changes the shape of the potential barrier and walls. Changing the heights of the potential barrier is precisely the manipulation of constraints. Thus, cells do thermodynamic work to build a structure called the membrane, which in turn manipulates constraints on reactions, meaning that cells do work at constructing constraints that manipulate constraints.

In addition, the cell does thermodynamic work to build an enzyme by linking amino acids together. It binds to the transition state that carries A and B to C and D—not to E or F and G—so it catalyzes that specific reaction, causing energies to reach down a specific pathway within a small number of degrees of freedom. You make C and D, but you don't make E and F and G. D may go over and attach to a transmembrane channel and give up some of its vibrational energy that popped the membrane open and allow in an ion, which then does something further in the cell. So cells do work to construct constraints, which then cause the release of energy in specific ways, so that work is done. That work then propagates, which is fascinating.

There are several points to keep in mind. One is that you cannot do a work cycle at equilibrium, meaning that the concept of an autonomous agent is inherently a non-equilibrium concept. A second is that once this concept is developed, it will only be a matter of 10, 15, or 20 years until, somewhere in the marriage between biology and nanotechnology, we'll make autonomous agents that will create chemical systems that reproduce themselves and do work cycles. So we have a technological revolution on our hands, because autonomous agents don't just sit and talk and pass information around; they can actually build things.

The third thing is that this may be an adequate definition of life. In the next 30 to 50 years, we're either going to make a novel life-form or we'll find one—on Mars, Titan, or somewhere else. I hope that what we find is radically different from life on Earth, because it will open up two major questions. First, what would it be like to have a general biology, a biology free from the constraints of terrestrial biology? And second, are there laws that govern biospheres across the universe? I'd like to think there are such laws. Of course, we don't know there are. We don't even know there are such laws for Earth's biosphere. But I have three or four candidate laws that I struggle with.

All of this points to the need for a theory of organization, and we can start to think about such a theory by critiquing the concept of work. If you ask a physicist what work is, he'll say that it's force acting through a distance. When you strike a hockey puck, for example, the more you accelerate it, the more little increments of force you've applied to it. The integral of that figure divided by the distance the puck has traveled is the work that you've done. The result is just a number.

In any specific case of work, however, there's an organization to the process. The description of the organization of the process that allows work to happen is missing from its numerical representation. In his book on the second law [of thermodynamics], Peter Atkins gives a definition of work that I find congenial. He says that work itself is a thing—the constrained release of energy. Think of a cylinder and a piston in an old steam engine. The steam pushes down on the piston, and it converts the randomness of the steam inside the head of the cylinder into the rectilinear motion of the piston down the cylinder. In this process, many degrees of freedom are translated into a few.

The puzzle becomes apparent when we ask some new questions. What are the constraints? Obviously the constraints are the cylinder and the piston, the fact that the piston is inside the cylinder, the fact that there's some grease between the piston and the cylinder so the steam can't escape, and some rods attached to the piston. But where did the constraints come from? In virtually every case, it takes work to make constraints. Somebody had to make the cylinder, somebody had to make the piston, somebody had to assemble them.

That it takes work to make constraints and it takes constraints to make work is an interesting cycle. This idea is nowhere to be found in our definition of work, but it's physically correct in most cases, and certainly in organisms. This means we are lacking theory, and points toward the importance of the organization of process.

The life cycle of a cell is simply amazing. It does work to construct constraints on the release of energy, which does work to construct more constraints on the release of energy, which does work to construct even more constraints on the release of energy, and other kinds of work as well. It builds structure. Cells don't just carry information. They build things, until something astonishing happens: A cell completes a closed nexus of work tasks and builds a copy of itself.

Although he didn't know about cells, Kant spoke about this 230 years ago, when he said that an organized being possesses a self-organizing propagating whole that can make more of itself. But although cells can do this, that fact is nowhere in our physics. It's not in our notion of matter, it's not in our notion of energy, it's not in our notion of information, and it's not in our notion of entropy. It's something else. It has to do with organization, propagation of organization, work, and constraint construction. All this has to be incorporated into some new theory of organization.

I can push this a little further by thinking of a puzzle about Maxwell's demon. Everybody knows about Maxwell's demon; he was supposed to separate fast molecules in one part of a partitioned box from the slow molecules by sending the slow molecules through a flap valve to another part of the box. From an equilibrium setting, the demon could then build up the temperature gradient, allowing work to be extracted. There's been a lot of good scientific work showing that at equilibrium the demon can never win. So let's go straight to a non-equilibrium setting and ask some new questions.

Think of a box with a partition and a flap valve. In the left side of the box there are *n* molecules and in the right side of the box there are *n* molecules, but the ones in the left side are moving faster than the ones in the right. The left side of the box is hotter, so there is a source of free energy. If you were to put a little windmill near the flap valve and open it, there would be a transient wind from the left to the right of the box, causing the windmill to orient itself toward the flap valve and spin. The system detects a source of free energy, the vane on the back of the windmill orients the windmill because of the transient wind, and then work is extracted. Physicists would say that the demon performs a measurement to detect the source of free energy. My new question is, How does the demon know what measurement to make?

Now the demon does a quite fantastic experiment. Using a magic camera, he takes a picture and measures the instantaneous position of all the molecules in the left and right [sides of the] box. That's fine, but from that heroic experiment the demon cannot deduce that the molecules are going faster in the left side than in the right side. If you took two pictures a second apart, or if you measured the momentum transfer to the walls, you could figure it out, but you can't with one picture. So how does the demon know what experiment to do? The answer is that the demon doesn't know what experiment to do.

Let's turn to the biosphere. If a random mutation happens by which some organism can detect and utilize some new source of free energy, and it's advantageous for the organism, natural selection will select it. The whole biosphere is a vast, linked web of work done to build things so that, stunningly enough, sunlight falls and redwood trees get built and become the homes of things that live in their bark. The complex web of the biosphere is a linked set of work tasks, constraint construction, and so on. Operating according to natural selection, the biosphere is able to do what Maxwell's demon can't do by himself. The biosphere is one of the most complex things we know in the universe, necessitating a theory of organization that describes what the biosphere is busy doing, how it is organized, how work is propagated, how constraints are built, and how new sources of free energy are detected. Currently we have no theory of it—none at all.

Right now I'm busy thinking about this important problem. The frustration I'm facing is that it's not clear how to build mathematical theories, so I have to talk about what Darwin called adaptations and then what he called pre-adaptations.

You might look at a heart and ask, What is its function? Darwin would answer that the function of the heart is to pump blood, and that's true—it's the cause for which the heart was selected. However, your heart also makes sounds, which is *not* the function of your heart. This leads to the easy but puzzling conclusion that the function of a part of an organism is a subset of its causal consequences, meaning that to analyze the function of a part of an organism, you need to know the whole organism and its environment. That's the easy part; there's an inalienable holism about organisms.

But here's the strange part: Darwin talked about pre-adaptations, by which he meant a causal consequence of a part of an organism that might turn out to be useful in some funny environment and therefore be selected. The story of Gertrude the flying squirrel illustrates this: About 63 million years ago, there was an incredibly ugly squirrel that had flaps of skin connecting her wrists to her ankles. She was so ugly that none of her squirrel colleagues would play or mate with her, so one day she was eating lunch all alone in a magnolia tree. There was an owl named Bertha in the neighboring pine tree, and Bertha took a look at Gertrude and thought, "Lunch!" and came flashing down out of the sunlight with her claws extended. Gertrude was very scared and she jumped out of the magnolia tree and, surprised, she flew! She escaped from the befuddled Bertha, landed, and became a heroine to her clan. She was married in a civil ceremony a month later to a handsome squirrel, and because the gene for the flaps of skin was Mendelian dominant, all their kids had the same flaps. That's roughly why we now have flying squirrels.

The question is, Could we have said ahead of time that Gertrude's flaps could function as wings? Well, maybe. Could we say that some molecular mutation in a bacterium allowing it to pick up calcium currents, thereby allowing it to detect a paramecium in its vicinity and escape the paramecium, could function as a paramecium-detector? No. Knowing what a Darwinian pre-adaptation is, do you think we could say ahead of time what all possible Darwinian pre-adaptations are? No, we can't. That means we don't know what the configuration space of the biosphere is.

It is important to note how strange this is. In statistical mechanics, we start with the famous liter volume of gas, and the molecules are bouncing back and forth, and it takes six numbers to specify the position and momentum of each particle. It's essential to begin by describing the set of all possible configurations and momenta of the gas, giving you a $6n$ dimensional phase space. You then divide it up into little $6n$-dimensional boxes and do statistical mechanics. But you begin by being able to say what the configuration space is. Can we do that for the biosphere?

I'm going to try two answers. Answer #1 is no, we don't know what Darwinian pre-adaptations are going to be—which supplies an arrow of time. The same is true of the economy: We can't say ahead of time what technological innovations will happen. Nobody was thinking of the Web 300 years ago. The Romans were using catapults to lob heavy rocks, but they certainly didn't have the idea of cruise missiles. So I don't think we can do it for the biosphere either, or for the econosphere.

You might say that it's just a classical phase space—leaving quantum mechanics out—and I suppose you can push me. You could say we can state the configuration space since it's simply a classical 6*n*-dimensional phase space. But we can't say what the macroscopic variables are, like wings, paramecium detectors, big brains, ears, hearing, and flight, and all the things that have come to exist in the biosphere.

All of this tells me that my tentative definition of an autonomous agent is a fruitful one, because it's led to all of these questions. I think I'm opening new scientific doors. The question of how the universe got complex is buried in this question about Maxwell's demon, for example, and how the biosphere got complex is buried in everything that I've said. We don't have answers to these questions; I'm not sure how to get answers. This leaves me appalled by my efforts, but the fact that I'm asking what I think are fruitful questions is why I'm happy with what I'm doing.

I can begin to imagine making models of how the universe gets more complex, but at the same time I'm hamstrung by the fact that I don't see how you can see ahead of time what the variables will be. You begin science by stating the configuration space. You know the variables, you know the laws, you know the forces, and the whole question is, How does the thing work in that space? If you can't see ahead of time what the variables are—the microscopic variables, for example— for the biosphere, how do you get started on the job of an integrated theory? I don't know how to do that. I understand what the paleontologists do, but they're dealing with the past. How do we get started on something where we could talk about the future of a biosphere?

There's a chance that there are general laws. I've thought about four of them. One of them says that autonomous agents have to live the most complex game they can. The second has to do with the construction of ecosystems. The third has to do with Per Bak's self-organized criticality in ecosystems. And the fourth concerns the idea of the adjacent possible. It just may be the case that biospheres on average keep expanding into the adjacent possible. By doing so, they increase the diversity of what can happen next. It may be that biospheres, as a secular trend, maximize the rate of exploration of the adjacent possible. If they did it too fast, they would destroy their own internal organization, so there may be internal gating mechanisms. This is why I call this an average secular trend, since they explore the adjacent possible as fast as they can get away with it. There's a lot of neat science to be done to unpack that, and I'm thinking about it.

One other problem concerns what I call the conditions of co-evolutionary assembly. Why should co-evolution work at all? Why doesn't it just wind up killing everything as everything juggles with everything and disrupts the ways of making a living that organisms have by the adaptiveness of other organisms? The same question applies to the economy. How can human beings assemble this increasing diversity and complexity of ways of making a living? Why does it work in the common law? Why does the common law stay a living body of law? There must be some very general conditions about co-evolutionary assembly. Notice that nobody is in charge of the evolution of the common law, the evolution of the biosphere, or the evolution of the econosphere. Somehow, systems get themselves to a position where they can carry out co-evolutionary assembly. That question isn't even on the books, but it's a profound question; it's not obvious that it should work at all. So I'm stuck.

Back Ad
TK